Environmental Sustainability and Development in Organizations

Challenges and New Strategies

Editors

Clara Inés Pardo Martínez

School of Administration
Universidad de Rosario
Bogotá, Colombia

Alexander Cotte Poveda

Faculty of Economy
Universidad Santo Tomas
Bogotá, Colombia

CRC Press is an imprint of the
Taylor & Francis Group, an **informa** business
A SCIENCE PUBLISHERS BOOK

First edition published 2021
by CRC Press
6000 Broken Sound Parkway NW, Suite 300, Boca Raton, FL 33487-2742

and by CRC Press
2 Park Square, Milton Park, Abingdon, Oxon, OX14 4RN

CRC Press is an imprint of Taylor & Francis Group, LLC

ISBN: 978-0-367-44086-2 (hbk) ISBN: 978-0-367-70727-9 (pbk)

Typeset in Times New Roman
by Radiant Productions

Foreword

As a scholar and leader in higher education and sustainability leadership, with deep interests in governance and organizations, I was delighted to be invited to introduce this important book. Having led a university as president for many years, I had the pleasure to advance learning and research in the field and develop strategic partnerships with organizations from the private and public sectors as well as work with social enterprises and non-governmental organizations. I also worked with national government departments in the UK and Europe to advance policy development. Now, as a senior researcher with Harvard University, I work with global chief executives and other senior leaders and collaborate with other learned faculty to advance transformational leadership for sustainable development across different business sectors and higher education, engaging with policy-makers at various levels.

Looking to address the challenges of environmental sustainability and development with new strategies relevant to organizations is an important and timely subject for academic discourse. This book explores different approaches to examine the relationship between environmental sustainability and development in public and private organizations with a focus on understanding how we can promote competitiveness and responsible growth. Given the breadth of the subject, the book has adopted the lens of sustainability as a catalyst for change as it relates to leadership and management, the role of technology, and how policy and other instruments can be used to transform sectors. The book illustrates new strategies in action as organizations are motivated to conserve and maintain natural resources, develop green supply chains, adopt low carbon approaches, and make efforts to deliver against the sustainable development goals through education.

Each chapter presents a detailed account of a particular topic and ranges widely across the field. The book places an emphasis on education and environmental literacy across sectors, highlighting the importance of values alongside knowledge and skills. An account of green economics and accounting draws attention to financial instruments and tax systems to advance green business. Drawing on competitiveness as a theme, the book offers a description of how pricing is no longer a sufficient strategy given more conscious consumers and regulations relating to sustainability. Offering deep insights into ecopreneurship and global entrepreneurship, the book examines sustainability in a sectoral context and shares detailed accounts from aviation, maritime, and urban development.

Distinguished by the concentration of accounts relating to the Global South and emerging economies, the book represents an important contribution to the field of environmental sustainability in organizational settings. By sharing insights into a range of new strategies relevant to a wide variety of public, private, and plural organizations, the book embodies a generous account of evidence-based approaches. I am pleased to commend the work of the authors and editors to those interested in the field of sustainability.

Wendy Maria Purcell
BSc PhD FRSA · Emeritus Professor of Biomedicine · Emeritus University President
Research Scholar with the T.H. Chan School of Public Health, Harvard University

Preface

It is important to publicize the future developments of environmental sustainability framed in the context of the different organizations in order to try to understand in a better way their impact on the development of emerging countries, Cotte, Pardo (2013). The vision of this text explores environmental issues from a long-term perspective, making various comparisons with the theoretical foundations built in recent years with the most innovative ideas and practices on the effect on economic development. The authors, from different perspectives, address from a first and updated vision some of the most outstanding issues in the different approaches of the theoretical application to the environment and its sustainability, bearing in mind the future and its conditions in a broad perspective and covering the different topics.

This book is structured to offer the most complete analysis possible of a wide range of questions that have been the subject of discussion within the research and study of environmental problems. It is of special interest to present an academic critique of what theory needs, practical application and about all the implementation of policies that contribute to the sustainability of the environment and natural resources. Several of the issues presented here are of vital importance to view their implications for development economics and to try to determine whether the near future can determine the optimal path to adequately manage economic growth with the environment so that it does not conflict with the quality of life that the various economies in their specific environment must experience.

The future must bear in mind the implications that organizations have on the quality of the environment, as well as its nature and scope, which are generators of a series of characteristics necessary for the construction of a theory in accordance with modern developments and that is a complement to the general theory to make the adaptations more consistent with the general welfare. Issues about environmental quality were considered for long periods of time as a problem solely and exclusively affecting industrialized economies; the technological advance that accompanied the industry in its beginnings was related to pollution, global warming, excessive dependence on certain types of chemical agents, and damage to natural resources, although it is true, these issues have been mitigated, the results are still not as desired.

The most recent studies and their evidence show that emerging countries today present serious environmental deterioration, the generalized belief of previous years affirmed that these countries exhibited fewer problems and very marginal effects on their environment since their technologies were more environmentally friendly. The most recent evidence affirms that there is intensive environmental degradation in developing countries with its implications on the quality of life, these findings demonstrate the importance of analyzing and conducting in-depth studies on the effects that exploitation causes on the environment, excessive resources use and how policies should be implemented to try to mitigate the damage caused.

The different chapters based on rigorous research have the main objective of carrying out a series of studies as a way of approaching the different measures that should be adopted in the context of the countries. It also deals with the different ways in which society's decisions affect the quality of the environment, the ways in which different resources are valued and the institutional implications that should require improvement in the quality of the environment and its sustainability, and most importantly, considers the design and implementation of more efficient policies to counteract the problems that are generated within the environment, using the most recent advances, mainly in the application and use of mechanisms and incentives to face environmental problems.

References

Cotte, A. and Pardo, C. I. 2013. CO_2 emissions in German, Swedish and Colombian manufacturing industries. Regional Environmental Change, 13, 979–988. https://doi.org/10.1007/s10113-013-0405-y.

Alexander Cotte Poveda
Faculty of Economy
Universidad Santo Tomas
Bogotá, Colombia

Clara Inés Pardo Martínez
School of Administration
Universidad de Rosario
Bogotá, Colombia

Contents

List of Reviewers

Wendy Purcell

BSc PhD FRSA · Emeritus Professor of Biomedicine · Emeritus University President Research Scholar with the T.H. Chan School of Public Health, Harvard University; wpurcell@hsph.harvard.edu

Nicoleta Cristache

Professor of Economics Science, University of Galati; nicoleta.cristache@ugal.ro

Siti Nurain Muhmad

Lecturer of Accounting, Universiti Malaysia Terengganu; sitinurain@umt.edu.my

Etika Ariyani

Universitas Muhammadiyah Mataram; ikachevy06@gmail.com

Diana Pinto Osorio

Professor, Universidad de La Costa; dianipin1@yahoo.es

Eugenia Buitrago

Pofessor, Unidad Central del Valle del Cauca; mbuitrago@uceva.edu.co

Ana María Ramirez

Universidad Autonoma de Occidente; aramirezt@uao.edu.co

Daniel Torralba

Score – Universidad del Rosario; daniel.torralba@urosario.edu.co

Arely Paredes

Conacyt, Facultad de Ciencias, UMDI Sisal; paredes.arely@gmail.com

Varsha Agarwal

Center for Management Studies Jain (Deemed-to-be University); varsha_a@cms.ac.in

He Zhu

Institute of Geographic Sciences and Natural Resources Research, Key Laboratory of Regional Sustainable Development Modeling, China; zhuhe@igsnrr.ac.cn

CHAPTER 1

Feminist Critical Discourse Analysis of Ecopreneurship as an Instrument for Sustainable Development

Grand Narratives and Local Stories

Anastasia-Alithia Seferiadis,[1,*] *Leah de Haan*[2] and *Sarah Cummings*[3]

Introduction

> *The ways in which social entrepreneurship travels globally as a concept and practice is riddled by problematic assumptions around 'Third-World' women's abilities and roles in the global economy. By demarcating the space in which particular women are legitimate entrepreneurial actors to microenterprise and social ventures, the field continues to exclude along gender lines, even if unintentionally. Such assumptions and practices can reproduce the ways in which women remain economically marginalized because of their confined legitimacy as founders and managers of 'less than' lucrative enterprises.*
>
> (Clark Muntean and Ozkazanc-Pan 2016: 223)

Sustainability studies, aiming to analyse potentially transformative models, have taken up the study of social enterprises operating in the ecology sector, known as ecopreneurship (Santini 2017). The widely used definition of Kirkwood and Walton (2010) defines ecopreneurs as 'entrepreneurs who start for-profit businesses with strong underlying green values and who sell green products and services' (2014). Given that non-profit organizations have been obliged to reduce their expectations in

[1] Laboratory Population Environment Development: Aix-Marseille University, Marseille, France and Institute for Research for Development, Montpellier, France.
[2] Chatham House, The Royal Institute of International Affairs, London, United Kingdom.
[3] Wageningen University, Wageningen, The Netherlands.
Emails: LdeHaan@chathamhouse.org; Sarah.Cummings@wur.nl
* Corresponding author: Anastasia.Seferiadis@gmail.com

terms of financing social activities through taxes and to generate more self-financing of their activities, social entrepreneurship—and herewith ecopreneurship—appear to provide a new framework for non-governmental development organizations and development 'beyond aid' (Fowler 2000). Neoliberal values of an independent, financial individual are apparently encapsulated in this approach (Wrenn 2015), linked to hegemonic masculinity (Ashe 2015). Given the increasing emphasis on the private sector within the current policy context of the Sustainable Development Goals (Cummings et al. 2019), social entrepreneurship is increasingly seen as a 'magic bullet' to solve development challenges by 'defying the obstacles that have prevented businesses from providing services to the poor' (Seelos and Mair 2005). In this context, ecopreneurship is perceived as a policy tool for solving the complex problems of the environmental and unemployment crisis (Isaak 2016). Ecopreneurship is used by researchers concerning ecopreneurs in different contexts, including women living in poverty (see, for example, Thamizoli et al. 2008).

The practices of ecopreneurship have often been described as particularly relevant for not only being a manner of sustainable development but also as potentially emancipatory trajectories (Altieri and Toledo 2011). However, around the world, social entrepreneurship, including ecopreneurship, largely consists of women questioning whether it is not simply another social reproduction mechanism that socialises women to give to others with many authors applying gendered analysis to study the social economy (Verschuur et al. 2015). Simultaneously, certain academic discourses on entrepreneurship have been shown to reinforce gender stereotypes (Ahl 2002). In order to provide insights into the capacity of ecopreneurship to contribute to the lives of women in the Global South, we question whether ecopreneurship is truly transformative in women's lives and hereby emancipatory from gendered discriminations.

In this chapter, we use feminist critical discourse analysis to interrogate the 'grand narrative' of ecopreneurship and examine it based on local stories of women ecopreneurs from Bangladesh, India and Ghana. After presenting how we use feminist critical discourse analysis to analyse ecopreneurship, building on Dey and Steyaert's conceptions of 'grand narratives', 'counter-narratives' and 'little narratives' (2010), we present how the ecopreneurship literature is constructing a grand narrative but also counter-narratives in particular based upon feminist scholarship. Next, building upon qualitative data collected between 2008 and 2019 from Bangladesh, India and Ghana, we analyse the little narratives of local women from the Global South. Finally, we discuss and conclude our analysis based on these results.

The practical value of this is dual. First, the approach highlights the hidden assumptions and tensions and their effect on praxis, in particular, as discourse analysis enables us to understand power relations and their effects (Foucault 1969). Second, it contributes to the work of feminist analyses for transformative sustainable development. This chapter is therefore useful for policymakers and practitioners in development, providing insights into local stories and how to counter inequalities through alternative discourses.

Feminist Critical Discourse Analysis

Discourse analysis is the collective name for a range of scientific methodologies for analysing how meaning is created and communicated through semiosis, comprising written, vocal or sign language. Discourse analysis is used in many disciplines in the social sciences, each with its own set of methodologies and assumptions. CDA is one form of the discourse analysis that aims to 'understand, expose, and ultimately resist social inequality' (van Dijk 2005). CDA focuses on the dialectical relationships between discourse and other elements of social practices. According to Fairclough (2012), networks of social practices constitute a social order and 'one aspect of this ordering is dominance: some ways of making meaning are dominant or mainstream in a particular order of discourse, others are marginal, or oppositional, or alternative' (Fairclough 2012).

Feminist CDA brings together CDA and feminist studies 'to show up the complex, subtle, and sometimes not so subtle, ways in which frequently taken-for-granted gendered assumptions and hegemonic power relations are discursively produced, sustained, negotiated, and challenged in different contexts and communities' (Lazar 2007). This is a very natural relationship as the establishment of CDA relied heavily on both gender studies and feminism and both are inherently critical of existing social structures. A feminist critical discourse analysis re-emphasises a crucial aspect of analysing sustainable development, namely its potential for social transformation. While CDA has an inherent affinity to emancipation, feminist theorising underscores the need for all social analyses to be about inducing change and thus sharpens CDA analyses. It demonstrates the need to go beyond critically analysing gendered inequalities and move towards prioritising true transformation.

In this feminist CDA, we employ Dey and Steyaert's (2010) conceptions of 'grand narratives', 'counter-narratives' and 'little narratives' to analyse the literature and practice of ecopreneurship. The grand narrative represents the dominant narrative which reflects patriarchy and other hierarchies of domination. The counter-narrative represents the critical feminist narrative because feminism particularly lends itself to pluralism and diversity. While there has been much criticism of ways in which the term 'feminism' has been employed as a transformative analytical tool is indeed intersectional. At its core, feminism concerns itself with a change that brings about gender equality and sees value, ownership and ability in all women. The only way this is possible is by including and deriving strength from the diversity that this includes women coming from all backgrounds, all sexualities and gender identities, all classes, all ethnicities and races and all abilities and all opinions.

Dey and Steyaert (2010) employ the term 'little narratives' to consider the 'little narratives of social inventiveness' (p. 97), counteracting the grand narrative in three ways. First, little narratives make the "social" visible. Second, they represent communal experiments at the limit of the grand narratives where 'it becomes possible to imagine novel subject positions and new forms of being' (p. 97). Third, they demonstrate the 'prosaic, unfinalisable character' of narratives (Dey and Steyaert 2010). Indeed, little narratives hold potential because they can acknowledge that social endeavours "necessarily exceed our capacity to know them" (Law 2004, Dey and Steyaert 2010). Although Dey and Steyaert (2010) were unable to go beyond some descriptions of potential little narratives, in the article we use descriptions of development projects conducted in Bangladesh, India and Ghana, derived from fieldwork undertaken by the first author since 2010. However, in order to move away from the exact binaries that feminist analyses attempt to destabilise, such as 'grand' versus 'little', we have chosen the term 'local stories'. This is in the knowledge that there is nothing 'grander' or more important to feminist analysis than the actual, local, lived experiences of actual women.

In this chapter, we employ transdisciplinary CDA (Fairclough 2012), amended by the authors in a series of earlier publications (Cummings et al. 2018, 2019, 2020), amended again by incorporating Dey and Steyaert's (2010) grand narratives, counter-narratives and little narratives and then further adapted to take a feminist perspective (see Table 1). We particularly rely on an earlier paper in which we described the CDA research process for a non-academic audience (Cummings et al. 2020). Transdisciplinary CDA traditionally comprises a four-phase research process, including a selection of a social question that can be productively approached by a focus on semiosis; identification of obstacles to addressing the social question based

Table 1. Methodology has been adapted from CDA and narrative analysis.

Critical discourse analysis	Narrative analysis	Feminist perspective	Approach in this article
Fairclough (2012), Cummings et al. 2020	Dey and Steyaert (2010)		Fairclough (2012), Dey and Steyaert (2010), Cummings et al. 2020
Phase 1: Identification of the social question			
Step 1: Selection of research topic that can be approached by focusing on text			Literature and stories of ecopreneurship
Step 2: Genealogy of past discourses	Grand narratives and counter-narratives	Dominant grand narratives and feminist counter-narratives	Dominant grand narratives and feminist counter-narratives
Phase 2: Selection and analysis of tests			
Step 1: Select appropriate texts	Little narratives		Creation of local stories based on fieldwork
Step 2: Analysis of the different texts			Analysis of the different texts
Step 3: Identify discourses in the text, based on past discourses identified in Phase 1			Identify discourses in the text, based on past discourses identified in Phase 1
Phase 3: Describe how the text was created			
Describe how the text was created			Describe how the local stories were created
Phase 4: Possible solutions and way forward			
Possible solutions or ways past the dominant discourse in terms of creating new discourses, narratives and arguments			Possible solutions or ways past the dominant discourse in terms of creating new discourses, narratives and arguments

Source: Authors

on the analysis of dialectical relations between semiosis and other social elements in texts; consideration of whether the social order needs the social question, namely whether it is inherent to the social order, whether it can be addressed within it or whether it can only be addressed by changing the social order; and identification of possible ways to past the obstacles with a semiotic point of entry through the use of discourses, narratives and arguments. In this chapter, the methodology has been adapted to incorporate Dey and Steyaert's (2010) conceptions of grand narratives, counter-narratives and local stories, making it more suitable to analysis of the literature and concrete examples of ecopreneurship. Given that feminism is concerned with action and social transformation, the fourth phase is concerned with changes to practice, as well as changes to discourse, narratives and arguments, already intrinsic to CDA.

In the next sections, we consider the grand narratives and feminist counter-narratives of ecopreneurship. This is followed by the local stories of ecopreneurship. In the discussion, we reflect on the local stories through the lens of the grand and the counter-narratives, followed by a reflection of the adequacy of the methodology.

Grand Narrative of Ecopreneurs: The Dominant Discourse

Analysis of the academic literature on ecopreneurship based on a literature review shows how a dominant discourse is being constructed.

Ecopreneurs, a term first coined by Isaak in 1998, are also known as 'sustainopreneurs' (Petersen and Schaltegger 2000), 'green entrepreneurs' and 'enviropreneurs' (Walley and Taylor 2002) 'sustainable entrepreneurs' (Santini 2017) and 'ecological entrepreneurs' (Rodríguez-García et al. 2019). They are characterised as demonstrating 'individual, self-driven, independent, highly motivated, environmental behaviour' (Pichel 2008). Walley and Taylor (2002) have developed a typology of four categories of ecopreneurs: innovative opportunists such as a fridge recycler, visionary champions such as a producer of natural skin and hair care products, ethical mavericks such as a craft exchange founder and ad hoc enviropreneurs producing, for example, organic pork. Ecopreneurs are often envisioned as being in a tension between profit and environmental ideals. Linnanen (2005) identifies four categories of ecopreneurs based on this tension between profit and environmental ideals, namely self-employed, non-profit business, opportunists and successful idealists. Research has also investigated the links between ecopreneurship and profitability (Porter and van der Linde 2000). Accordingly, venture capitalists and business angels are perceiving such competitive advantage which has led to more access to funds (O'Rourke 2016), which has accompanied an increased consumer demand for environmentally friendly products (Laroche et al. 2001) and has enabled environmental marketing, hereby enhancing companies' reputations (Miles and Covin 2000).

Ecopreneurs are not only innovative in terms of the products and services they develop but also their relationship to the market or the forms of enterprises adopted. Indeed, ecopreneurs are described as 'pioneers' (Lasner and Hamm 2011) as being as able to recognise opportunities (Keogh and Polonsky 1998) or as 'change drivers' (Walley and Taylor 2002) who can shape companies (Schaltegger 2002,

Kirkwood and Walton 2010). This demonstrates the dual role that ecopreneurs are seen as fulfilling: they are considered 'eco-conscious change agents' (Pastakia 1998) defined by both their motivation and their 'authenticity' (Santini 2017) with a 'less materialistic attitude' (Phillips 2005), and they have 'responsible' business and practices and values in opposition to traditional entrepreneurship (Schaper 2002). Not only are ecopreneurs described as value-laden but also that it is their belief set and motivations that shape firms (Walley and Taylor 2002).

Ecopreneurs are linked to a system of incubators, accelerators, investors, non-profit organisations, NGOs (non-governmental organisations), professional associations, higher education programmes (within business schools in particular), research institutions, forums and networks. Training and networking opportunities are provided. Competitions enable publicity and are accompanied by prize money. Successful role models are featured on websites, invited to testify at conferences and in embassies. During 'pitch days', ecopreneurs are given the opportunity to 'sell' their idea to private companies, intermediary organisations or international institutions. Funding is provided through grants, action funds and loans. Intermediary actors promote ecopreneurship specifically or via social entrepreneurship, supported by actors from the private sector or international institutions. Such a system aims to stimulate ecopreneurship because it has ostensibly high potential to solve complex problems, such as environmental problems, poverty and women's empowerment, while not requiring explicit state intervention and funding.

Ecopreneurs are also envisioned as contributing to changing the economic paradigm towards a green and sustainable economy (Pastakia 1998). As analysed by Santini (2017), the World Business Council for Sustainable Development describes ecopreneurs as change drivers who can foster a change from business as based on resource exploitation to resource preservation (page 6). The "visionary champion type of ecopreneur", as described by Walley and Taylor (2002), is "a champion of sustainability [who] sets out to change the world, operates at the leading edge, and has a vision of a sustainable future which has a hard, structural edge" (page 38). In this way, the ecopreneur is portrayed as a hero (Johnsen and Sørensen 2017). John Ogbor (2000) who applies discourse analysis to entrepreneurship shows that entrepreneurship discourse and praxis "reinforces an expression of patriarchy by producing and reproducing entrepreneurial ideas" which gives "primacy" to values considered masculine and therefore seen as belonging to men (page 626). The discourses develop the myth of the heroic entrepreneur, celebrating male concepts of control, competition, rationality or dominance. A study of discourses around female social entrepreneurship from a communications campaign in France shows this supposedly creates the 'superwoman', an exemplary version of individualised entrepreneurial femininity (Byrne et al. 2019) or the female version of the hero.

As mentioned above, ecopreneurship is increasingly seen as a path out of poverty. For example, 'ecopreneurship is on the rise in Africa, creating local solutions to alleviate poverty and various environmental problems ailing the different countries on the continent' (Dickens 2019). Ecopreneurship is seen as a way of simultaneously reducing the growing waste problem, providing employment to local people and supporting the transition to a more energy-intensive lifestyle without the use of fossil fuels (Dickens 2019). Other authors demonstrate that ecopreneurship can play a role in women's empowerment in developing countries (Maas et al. 2014).

The Feminist Counter-Narrative

Discussions of ecopreneurship often describe how women are more inclined towards ecopreneurship than men (Braun 2010). In fact, Rodgers (2008) coined the specific name 'shecopreneurship' to describe women ecopreneurs. Researchers have given many different reasons for this supposed affinity, namely that women are close to nature and socially inclined, the gendered division of labour, lack of a business mindset and superwoman as the new *homo economicus*. This section analyses counter-narratives from a feminist scholarship.

Close to Nature

First, some have concluded that women express higher levels of concern towards the environment (Zelezny et al. 2000, Davidson and Freudenberg 1996). Braun (2010) links it with women having 'stronger ethics of care' (using the concept of Chodorow 1974). Another study on women and ecopreneurship state that women's roles come from 'their intimacy with nature' and that they have a 'natural inclination towards the environment' (Potluri and Phani 2020), with the authors drawing from previous research which showed that women are 'spiritually close to nature' (Tøllefsen 2011). However, these results are based upon discourses without data on practices.

We can here see discourses that essentialise women's role in ecopreneurship build on the nature-culture dichotomy associated with femininity and masculinity. This can be linked to how ecofeminist thinkers and activists relate the domination and oppression of women with the domination and oppression of nature and link environment degradation with patriarchy (Mies and Shiva 1993). It should be noted, however, that because of the underlining ideas linking women with nature, ecofeminism can still be perceived as a form of essentialism. Critics stress that ecofeminism can restrict women to the natural world, reinforcing patriarchy's dichotomy of men and culture as opposed to women and nature (Davion 1994).

Socially Minded

Second, others, such as Huysentruyt (2014), consider that there is a smaller gender gap in social entrepreneurship when compared to traditional commercial entrepreneurship because women have a higher level of altruism and a stronger preference for redistribution, concluding that women are more socially minded and more averse to competition. These traits, identified by Huysentruyt (2014), are consistent with 'symbolic maternity' (Muel-Dreyfus 1996) which characterises women's work across cultures and throughout history, reinforcing gender subordination. Other commentators have likewise argued that women 'have a natural inclination to create organizations with social goals and intentions in mind' (Lortie et al. 2017). Lortie and colleagues (2017) collected survey data collected from 150 enterprises, showing that gender positively influences social salience which is measured as the establishment of social goals, the fulfilment of social needs, helping people other than customers and using resources for socially-oriented purposes. Indeed, it appears that women are overrepresented in all forms of the social economy, possibly sending women back to the realms of social reproduction or sacrifice for the well-being of their families.

This demonstrates what Susan Clark Muntean and Banu Ozkazanc-Pan (2016) detailed when analysing social entrepreneurship through several feminist perspectives, namely that the term social entrepreneurship is subject to complex gender connotations; if the term 'entrepreneur' is associated with the masculine (heroic, ambitious, courageous, strong and enterprising), the term 'social' denotes a female commitment (concerns related to exclusion, marginalisation, suffering, and creative activities, related to empathy). Eleanor Hamilton (2014) applies discourse and gender analysis to entrepreneurship and shows that despite a dominant discourse on male entrepreneurship, identities are 'are more fractured, open and contested than a categorization into masculine or feminine does not allow it' (p. 707), identities are indeed both contested and legitimised. Eleanor Hamilton explains that 'individuals define themselves in relation to each other and to the broader discourses available to them' (Hamilton 2014). As others have put forward, this shows how discourses produced by researchers reinforce gender stereotypes whereas it could be expected for researchers not to reproduce the domination of women through the production of knowledge but to show diversity (Clark Muntean and Ozkazanc-Pan 2016).

Gendered Division of Labour

Helene Ahl (2002) also shows that research discourses on entrepreneurship reinforce gender stereotypes regarding masculine forms of work and feminine forms of work. Not only is the concept of the entrepreneur constructed in a masculine way, but it also involves gender discrimination and in particular a gendered division of labour. In this division of labour, the entrepreneur is described as voluntary, determined, persistent, resolute, detached and self-centred, supposing that a woman performs unpaid and reproductive work associated with the private sphere.

This also raises questions of the appropriate indicators to measure social entrepreneurship and/or ecopreneurship. Indeed, as we have mentioned earlier, most tools use a monetarisation approach to attribute value to social or environmental impact. Taking the example of domestic work, Jany-Catrice and Meda (2013) also point out that monetarisation does not automatically make it possible to improve women's condition by giving value to their domestic activities, that the source of value is not in the work value but in the process of exchange and that the approach neglects the affective and relational character.

In addition to reproducing gender stereotypes on types of work, Janice Byrne, Salma Fattoum and Maria Cristina Diaz Garcia (2019) show that the 'role models' of women entrepreneurs mask racial, social class and gender barriers and standardise discriminatory treatment in the workplace. From this perspective, entrepreneurship is described as an appropriate alternative for working mothers as it allows them more apparent flexibility. As Susan Clark Muntean and Banu Ozkazanc-Pan (2016) explain, the flexibility of self-employment provided with social entrepreneurship, in contrast to paid employment, can create an additional burden for women who will earn less money and assume a larger share of traditional family responsibilities, thus reinforcing the sexual division of domestic work. Social entrepreneurship thus seems to constitute spaces where gender roles are reproduced rather than spaces of celebration where individuals engage in social innovations that change the world. In addition, in the social entrepreneurship sector, women are engaging in their business for their community. As the works of Blandine Destremau (2013) show, it is thus additional exploitation of women that takes place through their involvement in development projects.

Not Business-Minded

Researchers also explain the connection between ecopreneurship and women by suggesting that women lack self-confidence, start small and stay small or avoid innovation. As Helene Ahl's work shows (2015), discourses on social entrepreneurship can limit women to 'entrepreneurial ghettos', to use the term of Donald Bowen and Robert Hisrich (1986). This ghettoisation of women can also be linked to the 'threat of stereotype' (Steel and Aronson 1995); stereotypes can be sources of anxiety that weakens performance. Women remain economically marginalised because of their confined legitimacy as founders and managers of 'less than' lucrative businesses (Clark Muntean and Ozkazanc-Pan 2016).

Local Stories

Based upon field research conducted by the first author between 2008 and 2019 in Bangladesh (18 months fieldwork), India (3 weeks fieldwork) and Ghana (6 weeks fieldwork) are analysed to provide local stories of women.

Seeds Exchange in Rural Bangladesh

The first story concerns the NGO, Pride, based in Jessore, Bangladesh. Pride is a local NGO that aims to empower farmers in rural areas through sustainable development practices. It is an NGO that describes itself as a learning organisation that has institutionalised the participatory methodologies of action research, thus making it possible to involve local populations (Chambers 1994). The research focused on a project involving income-generating activities that target isolated villages that are not connected to the main roads, that are not accessible by bus, without shops in the villages and where no other NGO is active. These villages are also selected by the NGO according to the criteria of poverty. Thus, the villages included have more than 70% of the inhabitants who live with an average daily wage of 80 Takas (about 1 Euro per day). These villages are also characterised by a high population density (more than 650 inhabitants per square kilometre) and thus subject to strong pressure on natural resources. Gradually, the project focused on ways to alleviate women's poverty as women expressed an interest to participate, unlike men who said that they did not have free time to start something new.

Based on participatory processes, training organised by the NGO aimed to facilitate exchange and co-creation of knowledge, while also raising awareness on various issues holding women in poverty. The women in the project have progressively engaged in different income-generating activities, managing with what is at hand according to the 'bricolage' principle of entrepreneurs, using untapped or underutilised resources. Thus, their activities include market gardening and fruit growing by exploiting (small) free spaces (constructing pergolas above ponds, for example), rearing poultry in the backyard, fish farming in small ponds or embroidery and sewing. Women entrepreneurs benefit not only on their behalf (producing goods or earning money) but are also training their neighbours in these income-generating activities, sharing their knowledge and skills as well as stimulating spaces for participatory learning between women. In addition, these women help each other in their respective activities. For example, they help each other to prepare the soil of their garden. These women entrepreneurs have developed a model of economic action corresponding to social entrepreneurship.

Women reported they have access to more resources, especially that their families eat better in quantity and diversity. Not only are goods now produced for self-consumption (for example, vegetables, eggs and fish) but also the sale of products increases the money available at the household level (the average income of these activities is multiplied by 10 after one year in the project, according to a longitudinal study of 26 women). The women also report being now empowered with increased participation in decision-making in their families. Each trained woman (social entrepreneur) has an average of about 100 beneficiaries (according to NGO figures). By 2013, PRIDE had trained a total of 136 women in 136 different villages.

Women describe their exchanges as 'gifts', gifts which are then returned. They clearly express that they invest in their social fabric: if they help, they will be helped and reciprocity will be enacted. Women also express duty to help others, a responsibility. They have gained the capacity to help their community, so they must be doing so. They also are talking about gains in status, a strengthened recognition from their community which results into more support and more access to resources, including to justice as women are now invited in the *Shalish* meetings, namely a semi-formal conflict resolution structure at the village level which generally excludes the poorest and women.

Solar Engineers Trained in India

The second story considers the Barefoot College, based in Tilonia, Rajasthan, India. One of its projects is the Solar Mamma project which supports women to bring sustainable electricity to remote, inaccessible villages. As it says on their website, 'solar energy provides electricity and reduces carbon emissions, but we also see it as a catalyst to create employment, boost income and provide self-reliant solutions for village communities.'

Illiterate or semi-literate women are selected from developing countries in Asia, Africa, America and the Pacific Island, together with a partner organization at the grassroots. They are trained in rural Rajasthan for six months to be able to build, install, maintain and repair solar panels and other solar technologies. In doing so, the technology is demystified according to Solar Barefoot staff. The Indian Technical and Economic Cooperation, a division of the Ministry of External Affairs, funds training in Rajasthan. The equipment is paid for by the trainee's community, its government but also by its grassroot partners. Before women embark on this six months' training, they reach an agreement with their communities that the community will provide the so-called 'Solar Mama' with a small salary for maintenance of the solar technologies when they return home.

During their stay in India, they build together the material to electrify 80 houses, material that will be shipped when they return from India. While the women are in Rajasthan, the opportunity is taken to also complement their training with the

Enriche Programme which as mentioned on their website which 'supports women's aspiration to transform themselves and their communities. It provides them with the opportunity to embark on an empowerment journey; a journey to gain the confidence, skills and knowledge they need to reach their full potential as agents of sustainable change and to support their entrepreneurial aspirations.' Through discussions, but also making use of several tools such as drawing, women can discuss together gender-related topics.

In this project, women can be described as ecopreneurs. Unlike Pride where women were the ones developing an innovative manner of performing their exchanges, here women follow a path. The entrepreneurial idea of the Solar Mamas does not come from them, but they do engage in innovative paths. Such a path is taking them abroad to learning about a technology normally associated with lengthy education and more associated with male skills. In addition, when they return home, they are provided with a small allowance.

In this story, women talked about sacrifice; the sacrifice to be far away from their families for six months in a country where neither the food, language or climate are familiar, although the sacrifice is worth it. As one woman explained: "It is like they put us in a very narrow pipeline so we can learn, but when we get out, we will learn something that is worth it" because they will then be able to provide their community with most needed electricity. I talked to these women while they were still in Rajasthan, being trained for over four months, and they did not talk to about earning an income from this activity when they get back to their village, although a small salary has been negotiated in advance with each community. Instead, what was prevalent in their narratives was their 'responsibility' towards their community.

Winners of a Social Entrepreneurship Programme in Ghana

The third story concerns ecopreneurs who participated in a social entrepreneurship competition in Ghana. These enterprises are led by young, university-educated Ghanaians. They have won a prize, are featured on websites, have been interviewed by the press and have been invited to present their enterprise at conferences in schools and in embassies. The women's enterprises are very different; some are about using abundant and underutilised resources that can be picked, and it is innovative economic exchanges such as barter or building upon other economic programmes, such as rotating credit schemes. It can include technologies such as solar panels but in partnerships with companies. The women interviewed explained how they discovered untapped opportunities but also how difficult it is to make profits. Some enterprises have never started, others are 'on hold' with many ecopreneurs employed in companies or in the development sector to earn an income. The women were also worried about engaging in partnership, to sell their idea and have the beneficiaries not protected, the 'vision not understood', the 'impact not a priority'. It is also women who express they want to change the narratives on women in development in Africa. For example, one interviewee said: "I work with women in rural areas. It sounds cliché, as if every woman works with the empowerment of women. And people put you aside". She, therefore, explains that she is being restricted to a certain area where her activities have a perceived legitimacy.

Discussion

The grand narratives illustrate ecopreneurship to be a development strategy whose objectives and considerations fit the neoliberal understanding of the world perfectly. They see the global economy as a meritocracy where all you require is to be *homo economicus*. Helene Ahl (2002) shows that the discourses of entrepreneurship, which construct the concept as something positive, associated with innovation, growth and development, are part of the great narrative of modernity where development is 'progress'. Development, thus, becomes the process where a person goes through an incubator, becomes economically awakened, can suddenly reap the benefits of the system while simultaneously receiving the responsibility to maintain it and, finally, is asked to either demonstrate this or at least affirm it. Dey and Steyaert (2010) show how such narratives on social entrepreneurship have in common utopian rhetoric and an emphasis on novelty. These narratives often contain a founding plot that legitimises a necessary break with the past as well as a discourse on performativity, rationalism, progress and individualism, all terms which are associated with masculinity. This not only generates a depoliticisation of social change, the exact opposite of feminist transformations but also either disregards gender or sees 'superwomen' as masculinised women who take on neoliberal, male characteristics to succeed in the global economy.

The definition of ecopreneurship in the grand narrative, focusing on the individual, self-driven, independent, highly motivated and environmental behaviours, does not appear to be consistent with the experience of female ecopreneurs in the local stories. Indeed, our local stories show women innovating more in the types of exchanges or seizing opportunities rather than innovating novel sustainability ideas. The women are not heroic in terms of the *homo economicus* but rather in their struggles to improve their own and their communities' lives. Indeed, this community aspect, mentioned in the Bangladeshi and India stories, appears to be largely missing in the grand narratives which are very much focused on the role of the individual entrepreneur. Feminist counter-discourses resist gender stereotyping which describes women as close to nature, socially inclined, assigned to a specific division of labour, not business minded, or as 'superwoman' who adopt the *homo economicus* stereotypes. Although feminist discourses provide criticism, this does not prevent women from being restricted to entrepreneurial ghettos without profits, which correspond to our local stories that show women making little or no profit instead of putting forward their ideas of responsibility.

The local stories show how women carefully re-craft social relations, how these social relations contribute by conferring a higher status and hence symbolic capital to activities of care, both social care and environmental care. As these stories show, women are not heroes who break from the dominant pattern instead they act within social frames and norms to strengthen interdependent ties within communities and with the environment. Feminist research needs to evaluate how responsibility towards the society and the environment can be enacted upon as a shared and emancipatory responsibility so that neither women nor nature are exploited.

Dey and Steyaert (2010) highlight the need to break free from the grand narratives and to focus on the 'little narratives' to explore the ambivalences and paradoxes of ecopreneurship. These little narratives, which we labelled local stories, helped us to make visible the power games and social hierarchies behind ecopreneurship. For Fowler (2000), social entrepreneurs—and implicitly ecopreneurs—are located within the market. For this reason, he questions whether 'civic innovators', actors emphasising popular engagement rather than business, can be successful within the market because they are not necessarily looking for market-led solutions and are looking for social solutions. For this reason, if promoted, feminist counter-discourses and the discourses evident in the local stories could be used to find different types of solutions for sustainability.

Conclusions

In the context of the trinity of women, the environment and sustainable development, it is necessary to engage with hierarchical and dominating structures, including patriarchy, capitalism, racism and classism. As we have shown, ecopreneurship cannot be considered an approach to development that confronts these power relations. The easy equation of women and nature, the small scale, the reliance on neoliberal conceptions of people and the focus on caring all reify these structures. That, however, is not to say that the women in the local stories we have shared do this. In fact, in the face of all the remaking that ecopreneurship does, they have found, created and broken open spaces for themselves. Thus, if any approach to sustainable development, focused on women and/or the environment, is to be truly transformative in the way that feminism intends, it will need to learn from the tactics, skills and originality women have had to develop to create these spaces. It is here where we can identify transformative, feminist action and it is with these stories and approaches in mind that the grand narrative needs to be addressed. Feminism is always wholly active and wholly theoretical, making it a perfect example of praxis. Thus, it should come as no surprise that the most perfect example of an alternative feminist discourse we can locate within ecopreneurship of feminism is in the way individual women conduct themselves in their situations.

This chapter sheds light upon examples of little narratives by analysing stories of local women in the Global South. However, this work remains limited and does not account for the plurality of how women across contexts innovate. To further the understanding of how ecopreneurship can be a transformative mode of action and therefore more data as well as context-based analysis of local women appropriation stories are needed.

Despite the manner in which the individualisation of ecopreneurship feeds the existing discourses in structures, it does allow space for the local feminist interventions of women by countering the 'one-size-fits-all' rhetoric. Ecopreneurship, though largely based on non-feminist rationales, does act as a counterweight to more macro-economic approaches to women's economic engagements. By focusing on this aspect of ecopreneurship, it is possible to see how localised, alternative discourses are given space to be enacted. This still requires women to innovate these forms of resistance and still places the larger care burden on women, but the lack of strict mould means there is at least some room for manoeuvre. In these different spaces, a myriad of different approaches has been developed, and it is this plurality that needs to be fostered to provide space for alternative, feminist discourses; an example of the plurality that feminism is inherently comfortable with.

For ecopreneurship, as a novel mode of economic activity, to generate transformative change that breaks from the exploitation of nature and of women and to become an example of ecofeminism (Agarwal 2002), there needs to be critical analysis the centrality of caring ties and non-hierarchical relations. If the social economy field can give value and primacy to reciprocal exchanges (Guérin 2003), it is these types of ties that need to be analysed with a non-compromising feminist perspective. If there is to be value in caring for one another and nature, this ought to be conceived in a non-hierarchical manner to transform power relations.

Acknowledgements

Anastasia-Alithia Seferiadis received funding to carry the fieldwork in Bangladesh by the Athena Institue, Vrije Universiteit Amsterdam, The Netherlands, in India by the Groupement d'intérêt scientifique de l'Institut du Genre (Gis Genre), Paris, France, and in Ghana by the Fondation de la Croix Rouge Française. Leah de Haan's contribution was not undertaken as part of Chatham House's research programmes or projects. Sarah Cummings' contribution to this research has been undertaken as part of the NWO-WOTRO NL-CGIAR project on 'Improving the effectiveness of public-private partnerships within the CGIAR: knowledge sharing for learning and impact' (W 08.240.301). The authors would also like to acknowledge with gratitude the role played by the anonymous reviewer.

References

Agarwal, B. 2002. Le débat sur le genre. Enjeux Contemporains du Féminisme Indien, 181, 155.

Ahl, H. J. 2002. The making of the female entrepreneur: A discourse analysis of research texts on women's entrepreneurship Doctoral dissertation. InternationellaHandelshögskolan.

Altieri, M. A. and Toledo, V. M. 2011. The agroecological revolution in Latin America: rescuing nature, ensuring food sovereignty and empowering peasants. Journal of Peasant Studies, 38(3), 587–612.

Ashe, P. 2015. Hegemonic masculinity and profeminism: using internarrative identity and intersectionality to move beyond neoliberal imperialism. *In*: Way, L. (ed.). Representations of Internarrative Identity. Palgrave Macmillan: London, United Kingdom.

Bowen, D. D. and Hisrich, R. D. 1986. The female entrepreneur: A career development perspective. Academy of Management Review, 112, 393–407.

Braun, P. and McEachern, S. 2010. Climate change and regional communities: Towards sustainable community behaviour in Ballarat. Australasian Journal of Regional Studies, The, 161, 3.

Byrne, J., Fattoum, S. and Diaz Garcia, M. C. 2019. Role models and women entrepreneurs: Entrepreneurial superwoman has her say. Journal of Small Business Management, 57(1), 154–184.

Chambers, R. 1994. The origins and practice of participatory rural appraisal. World Development, 227, 953–969.

Chodorow, Nancy. 1974. Family structure and feminine personality. pp. 43–66. *In*: Joan Bamberger, Louise Lamphere and Michael Zimbalist Rosaldo (eds.). Women, Culture and Society. Stanford: Stanford University Press.

Chodorow, N. 2001. Family structure and feminine personality. Feminism in the Study of Religion, 81–105.
Clark Muntean, S. and Ozkazanc-Pan, B. 2016. Feminist perspectives on social entrepreneurship: critique and new directions. International Journal of Gender and Entrepreneurship, 83, 221–241.
Cummings, S., Regeer, B., de Haan, L., Zweekhorst, M. and Bunders, J. 2018. Critical discourse analysis of perspectives on knowledge and the knowledge society within the sustainable development goals. Development Policy Review, 366, 727–742.
Cummings, S., Seferiadis, A. A. and de Haan, L. 2019. Getting down to business? Critical discourse analysis of perspectives on the private sector in sustainable development. Sustainable Development. https://doi.org/10.1002/sd.2026.
Cummings, S., De Haan, L. and Seferiadis, A. A. 2020. How to use critical discourse analysis for policy analysis: a guideline for policymakers and other professionals. Knowledge Management for Development Journal, 15(1), 99–108.
Davidson, D. J. and Freudenburg, W. R. 1996. Gender and environmental risk concerns: A review and analysis of available research. Environment and Behavior, 283, 302–339.
Davion, V. 1994. Is ecofeminism feminist? Ecological Feminism, 8–28.
De Bruin, A. 2016. Towards a framework for understanding transitional green entrepreneurship. Small Enterprise Research, 23(1), 10–21, DOI: 10.1080/13215906.2016.1188715.
Destremau, B. 2013. Au four, au moulin... et à l'empowerment. La triple captation et l'exploitation du travail des femmes dans le développement. Travail et genre dans le monde. L'état des savoirs.
Dey, P. and Steyaert, C. 2010. The politics of narrating social entrepreneurship. Journal of Enterprising Communities: People and Places in the Global Economy.
Dickens, A. 2019. How ecopreneurs alleviate poverty in Africa. Borgen Magazine. 25 May 2019. https://www.borgenmagazine.com/how-ecopreneurs-alleviate-poverty-in-africa/. Accessed 25 February 2020.
Dixon, S. E. and Clifford, A. 2007. Ecopreneurship—a new approach to managing the triple bottom line. Journal of Organizational Change Management, 20(3), 326–345.
Fairclough, N. 2012. Critical discourse analysis. pp. 9–21. *In*: Gee, J. P. and Handford, M. (eds.). The Routledge Handbook of Discourse Analysis. Routledge: Abingdon, United Kingdom and New York, USA.
Foucault, M. 1969. L'Archéologie du savoir, Paris, Gallimard. Bibliothèque des sciences humaines.
Fowler, A. and Mati, J. M. 2019. African gifting: pluralising the concept of philanthropy. VOLUNTAS: International Journal of Voluntary and Nonprofit Organizations, 1–14.
Fowler, A. 2000. NGDOs as a moment in history: beyond aid to social entrepreneurship or civic innovation? Third World Quarterly, 21(4), 637–654.
Gibbs, D. 2009. Sustainability entrepreneurs, ecopreneurs and the development of a sustainable economy. Greener Management International, 55.
Gilligan, C. 1993. In a Different Voice: Psychological Theory and Women's Development. Harvard University Press: Cambridge, USA.
Guérin, I. 2003. Economie solidaire et inégalités de genre: une approche en termes de justice sociale. Revue internationale de l'économie sociale: recma, 289, 40–56.
Hamilton, E. 2014. Entrepreneurial narrative identity and gender: a double epistemological shift. Journal of Small Business Management, 524, 703–712.
Huysentruyt, M. 2014. Women's social entrepreneurship and innovation. OECD Local Economic and Employment Development (LEED) Papers. https://doi.org/10.1787/20794797.

Isaak, R. 2016. Ecopreneurship, rent-seeking, and free-riding in global context: Job-creation without ecocide. Small Enterprise Research, 231, 85–93.

Johnsen, C. G. and Sørensen, B. M. 2017. Traversing the fantasy of the heroic entrepreneur. International Journal of Entrepreneurial Behavior and Research, 232, 228–244.

Keogh, P. D. and Polonsky, M. J. 1998. Environmental commitment: a basis for environmental entrepreneurship? Journal of Organizational Change Management, 111, 38–49.

Kirkwood, J. and Walton, S. 2010. What motivates ecopreneurs to start businesses? International Journal of Entrepreneurial Behavior and Research.

Kolawole, O. D. and Torimiro, D. O. 2005. Participatory rural entrepreneurship development for grassroots transformation: a factor analysis. Journal of Human Ecology, 18(3), 193–198.

Laroche, M., Bergeron, J. and Barbaro-Forleo, G. 2001. Targeting consumers who are willing to pay more for environmentally friendly products. Journal of Consumer Marketing, 186, 503–520.

Lasner, T. and Hamm, U. 2011. Ecopreneurship in aquaculture—The adoption of organic fish farming methods. Organic Is Life Knowledge for Tomorrow, 2, 72–75.

Law, J. 2004. After Method: Mess in Social Science Research. Routledge, New York, NY.

Lazar, M. M. 2007. Feminist critical discourse analysis: articulating a feminist discourse praxis. Critical Discourse Studies, 4(2), 141–164, DOI: 10.1080/17405900701464816.

Linnanen, L. 2005. An insider's experiences with environmental entrepreneurship. Making Ecopreneurs: Developing Sustainable Entrepreneurship, 72–88.

Lortie, J., Castrogiovanni, G. J. and Cox, K. C. 2017. Gender, social salience, and social performance: how women pursue and perform in social ventures. Entrepreneurship & Regional Development, 29(1-2), 155–173.

Maas, J., Seferiadis, A. A., Bunders, J. F. and Zweekhorst, M. B. 2014. Bridging the disconnect: How network creation facilitates female Bangladeshi entrepreneurship. International Entrepreneurship and Management Journal, 10(3), 457–470.

Mies, M. and Shiva, V. 1993. Ecofeminism. Zed Books.

Mies, M. and Vandana, S. 1993. Ecofeminism. Halifax, N.S.: Fernwood Publications, 24.

Miles, M. P. and Covin, J. G. 2000. Environmental marketing: A source of reputational, competitive, and financial advantage. Journal of Business Ethics, 233, 299–311.

Muel-Dreyfus, F. 1996. Vichy et l'éternel féminin: contribution à une sociologie politique de l'ordre des corps. Seuil.

Nicholls, A. 2009. 'We do good things, don't we?': 'Blended value accounting' in social entrepreneurship. Accounting, Organizations and Society, 346-7, 755–769.

Ogbor, J. O. 2000. Mythicizing and reification in entrepreneurial discourse: Ideology-critique of entrepreneurial studies. Journal of Management Studies, 37(5), 605–635.

O'Rourke, A. R. 2016. How venture capital can help build ecopreneurship. pp. 185–204. *In*: Michael Schaper (ed.). Making Ecopreneurs: Developing Sustainable Entrepreneurship. Routledge: Abingdon, United Kingdom, and New York, United States, 165.

Palmas, K. and Lindberg, J. 2013. Livelihoods or ecopreneurship? Agro-economic experiments in Hambantota, Sri Lanka. Journal of Enterprising Communities: People and Places in the Global Economy, 7.2, 125–135.

Pastakia, A. 1998. Grassroots ecopreneurs: change agents for a sustainable society. Journal of Organizational Change Management.

Phillips, M. 2005. June. Ecopreneurs making green sense: Reflections on two case studies. pp. 13–15. In Proceedings of the British Academy of Management Conference, Oxford, UK.

Pichel, K. 2008. Enhancing ecopreneurship through an environmental management system: A longitudinal analysis of factors leading to proactive employee behaviour. Sustainable Innovation and Entrepreneurship, 141–182. DOI: 10.4337/9781848441552.00015.

Porter, M. and Linde, C. 2000. Green and competitive: ending the stalemate. The Dynamics of the Eco-Efficient Economy, Edward Elgar, Cheltenham/Northampton, 33–55.

Potluri, S. and Phani, B. V. 2020. Women and green entrepreneurship: a literature based study of India. International Journal of Indian Culture and Business Management, 20(3), 409–428.

Rodgers, C. and Director, D. B. A. 2008. "Shecopreneurship:" Female ecopreneurs and how they do business. In Sustainable Innovation 08: Future Products, Technologies and Industries.

Rodríguez-García, M., Guijarro-García, M. and Carrilero-Castillo, A. 2019. An overview of ecopreneurship, eco-innovation, and the ecological sector. Sustainability, 1110, 2909; https://doi.org/10.3390/su11102909.

Santini, C. 2017. Ecopreneurship and ecopreneurs: Limits, trends and characteristics. Sustainability, 9(4), 492.

Schaltegger, S. 2002. A framework for ecopreneurship. Greener Management International, 38.

Schaper, M. 2002. The challenge of environmental responsibility and sustainable development: Implications for SME and entrepreneurship academics. Radical Changes in the World: Will SMEs Soar or Crash, 541–553.

Seelos, C. and Mair, J. 2005. Social entrepreneurship: Creating new business models to serve the poor. Business Horizons, 483, 241–246.

Steele, C. M. and Aronson, J. 1995. Stereotype threat and the intellectual test performance of African Americans. Journal of Personality and Social Psychology, 695, 797.

Thamizoli, R. R. P., Balasubramanian, K., Selvamukilan, B., Devaraj, M., Nair, S., Alluri, K. and Shanmuganathan, M. R. 2008. ODL for Ecopreneurship: Promotion of Multiple Livelihoods among the Women SHGs in Tamil Nadu, India. https://www.researchgate.net/profile/RENGALAKSHMI_Raj/publication/280722919_ODL_for_Ecopreneurship_Promotion_of_Multiple_Livelihoods_among_the_Women_SHGs_in_Tamil_Nadu_India/links/55c2fbf508aea2d9bdbff511/ODL-for-Ecopreneurship-Promotion-of-Multiple-Livelihoods-among-the-Women-SHGs-in-Tamil-Nadu-India.pdf. Accessed 24 February 2020.

Thompson, N., Kiefer, K. and York, J. G. 2011. Distinctions not dichotomies: Exploring social, sustainable, and environmental entrepreneurship. pp. 201–229. *In*: Lumpkin, G. T. and Katz, J. A. (ed.). Social and Sustainable Entrepreneurship (Advances in Entrepreneurship, Firm Emergence and Growth, Vol. 13). Emerald Group Publishing Limited, Bingley.

Tøllefsen, I. B. 2011. Ecofeminism, religion and nature in an Indian and global perspective. Alternative Spirituality and Religion Review, 21, 89–95.

Van Dijk, T. A. 2005. Critical discourse analysis. *In*: Schiffrin, D., Tannen, D. and Hamilton, H.E. (eds.). The Handbook of Discourse Analysis. Blackwell Publishers: Malden, Massachusetts, USA.

Verschuur, C., Guérin, I. and Hillenkamp, I. 2015. Pourquoi croiser l'économie féministe et l'économie sociale et solidaire? pp. 21–28. *In*: Verschuur, C., Guérin, I. and Hillenkamp, I. (eds.). Homo oeconomicus, muliersolidaria. Une économie solidaire peut-elle être féministe? L'Harmattan: Paris, France.

Walley, E. E. and Taylor, D. W. 2002. Opportunists, champions, mavericks...? Greener Management International, 38.

Wrenn, M. V. 2015. Agency and neoliberalism. Cambridge Journal of Economics, 39(5), 1231–1243. September 2015, https://doi.org/10.1093/cje/beu047.

Zelezny, L. C., Chua, P. P. and Aldrich, C. 2000. New ways of thinking about environmentalism: Elaborating on gender differences in environmentalism. Journal of Social Issues, 563, 443–457.

Chapter 2

Environmental and Natural Resources Research

The Case of a Research Institute in Colombia

Alexander Cotte Poveda[1,*] and *Clara Inés Pardo Martínez*[2,*]

Introduction

The research process has changed in recent decades due to new technologies, communication strategies, peer review, open science, among other factors, generating multinational endeavours, international collaborations and financing using efficient research resources and strengthening scientific products to prevent research misconduct (OECD 2009). These elements are important to determine the relationship between national and international policies that allow us to achieve pertinent and applicable knowledge by adopting good practices in research to promote sustainable development and welfare of the population.

Currently, research is highly competitive because of different stakeholder pressures and requirements for the outcomes of the successful quest for new knowledge, which determine financing, ranking positions, prestige, new projects, changes in theories, influence in policy decisions, etc. (ESF 2010). For these reasons, it is necessary to build and develop trust and efficient relationships among science, society and policymakers that guarantee the culture of best practices and determine the scientific production that contributes to the solution of global problems.

Research institutes play an important role in science, technology and innovation activities due to their inter- and intraregional networks and collaboration, which are recognised as knowledge providers, diffusing and transferring knowledge and information from science and technology to business, helping access internationally

[1] Universidad Santo Tomas, Faculty of Economy, Bogotá, Colombia.
[2] Universidad de Rosario, School of Administration, Bogotá, Colombia.
* Corresponding author: alexcotte@yahoo.com; cipmusa@yahoo.com

available sources of knowledge, providing a highly skilled workforce and supporting businesses in their proximity (Fritsch and Schwirten 1998, Charles and Goddard 1997). In this context, the importance of research institutes is as follows (Revilla 2000):

i. Research institutes support the science, technology and innovation (STI) process of business, especially in small and medium enterprises.
ii. These institutions promote growth and development with the generation of new products, services and processes to promote efficiency and competitiveness.
iii. Research institutes support and assist local organisations in their STI efforts and promote entrepreneurship and the viability of the business.
iv. These organisations cooperate and collaborate with other institutions outside the region, extending the knowledge base of the territory.
v. The results of research institutes serve as inputs to policymakers and decision-makers to support evidence-based strategies and policies. These issues demonstrate the importance of research institutes and the importance of evaluating their results based on the scientific products generated.

Research institutes are quite diverse and interact within and among countries according to knowledge. Their activities and processes vary widely according to their mission, type, financing, structure and aims. Some are perceived as having high risks with uncertain returns when dynamics and issues in the long term are considered, while others emphasise more short-term, market-oriented research, development work, problem-solving, technical assistance, applied knowledge and supporting evidence (OECD 2011), which is fundamental to understanding scientific production and the role of science according to stakeholders or function in society.

Developed countries have achieved models to promote STI through research institutes by determining key factors that influence success and failure, such as the following (Intarakumnerd and Goto 2016):

i. Funding should come from different stakeholders but especially from industry and the government to guarantee the achievement of its mission with the resources required to operate.
ii. Researchers are important for maintaining research capability and quality when research issues change over time to achieve this objective. Research institutes need to constantly upgrade research capability by employing new researchers with new capabilities and interests and to provide core researchers to ensure the permanency of research and operation of institutionalised organisational knowledge.
iii. Setting a research agenda from internal and external analysis takes into account customer demand and requirements of government that bridge the gap between 'basic' and 'applied' research by both obtaining a fundamental understanding of scientific problems and, at the same time, seeking to be eventually beneficial to society (Stokes 1997).
iv. Performance evaluation is fundamental for guaranteeing results and reputation and obtaining financing from industries or grants and other incentives. For

this reason, research institutes must design metrics and indicators to analyse performance and potential improvements in processes and scientific production.

v. Geographical matters allow research institutes to function effectively as knowledge hubs; geographical proximity allows researchers to share facilities, maintain close contacts, conduct collaborative research, receive technological advice and supervise PhD and post-doctoral researchers.

vi. Governance must be balanced between inputs from industry on the management and governance of research institutes and appointing distinguished researchers from academia to structure the overall strategic direction of these organisations and the directions of research programs and promotion of internal staff. These elements show the importance of the structure and mission of research institutes to mitigate STI failures among businesses and non-firms through mechanisms, such as research and development consortiums, specialised programmes for industries and SMEs and connections between STI, experts, governments and the productive sectors in different geographical regions.

Research institutes require effective management activities that include (OECD 2014) attracting funding, managing funds, liaising with funding bodies, project planning, developing research projects, generating new knowledge and implementing, monitoring and evaluating research outcomes. All this comes on top of research activities, such as producing publications, disseminating research, gaining patents, developing new processes, producing goods and services and in many cases commercialising research. These activities generate technology-based businesses and entrepreneurship, which is important for achieving sustainable development and knowledge-based society.

Scientific production is generally measured as the number of papers published in leading peer-reviewed journals in indexed databases (Scopus and Clarivate). However, the volume of scientific output differs and depends on the institutionalised structures of higher education and research, and increasing science productivity demands commensurate resources according to different studies that have confirmed the positive relationship between research funding and publication output which implies the conditions necessary for continuous and strong growth in scientific productivity (Weingart 2015, Rosenbloom et al. 2014, Powell and Dusdal 2017). This demonstrates the importance of evaluating scientific performance through the number of papers and other indicators.

With this background and taking into account the importance of research institutes to promote sustainable development, this study sought to analyse scientific performance in the environment and natural resources of a research institute in an emerging economy with such high biodiversity as Colombia through bibliometric analysis to compare the results of research products and to determine the impact and pertinence of these results.

The chapter is set out as described below. Section 2 examines the methods used in this study based on a case study, and Section 3 discusses the results of trends and dynamics of research on the environment and natural research in the research institute. Finally, conclusions are highlighted in Section 4.

Methods

This study used different scientometric and bibliometric methods to determine the trends and dynamics of scientific research in the environment and natural resources in a research institute in a developing country, Colombia.

The different trends and evolution in the research on environmental and natural resources, which is one of the institutes' central research focuses, were determined. To achieve this, the different indicators resulting from the Scopus author identifier were used. This tool assigns a unique number to groups of documents written by the same author through an algorithm that in technical and practical terms coincides with the authorship of most documents. Publications were found in the database, and for this case of analysis, they were included based on certain classification criteria. Thus, for example, if a document could not be definitively linked with an author identifier, this could cause more than one entry to exist for the same author. In terms of aggregation by author, this implies that duplication has effects on the total number of documents in the citation indexes and therefore in individual and institutional authorship. For this reason, if the authorship was not identified with full evidence, it was grouped separately.

For the different interrelationships by research topics, a relationship was modelled for international institutions with their areas, their topics and the diversity of production according to the speciality interest.

Results and Discussion

This section shows the main results of the research products in the research institute in Colombia. Table 1 shows by country the evaluation of publications by country at the Latin American level in the area of environmental sciences. For the sample of 10 representative countries, Brazil was first in the ranking by a number of documents published and a number of citations followed by Mexico. Colombia appears in fifth place with 1,288 documents published in the 1996–2018 period and a total of 24,137 citations.

Table 1. Documents published in environmental sciences Latin America 1996–2018.

Rank	Country	Documents	Documents cited	Citations	Auto-citations	Citations by document	H index
1	Brazil	12,696	12,292	200,051	84,451	15.76	152
2	Mexico	6,735	6,589	108,219	27,447	16.07	117
3	Argentina	5,618	5,494	102,014	34,683	18.16	111
4	Chile	3,147	3,069	55,994	15,417	17.79	86
5	Colombia	1,288	1,246	24,137	3,014	18.74	70
6	Panama	762	739	26,729	2,690	35.08	85
7	Ecuador	715	698	12,526	1,664	17.52	55
8	Costa Rica	642	627	14,928	1,478	23.25	60
9	Peru	638	622	13,990	1,751	21.93	58
10	Puerto Rico	554	539	20,107	1,165	36.29	68

Source: Scimago

Table 2 determines in order of published documents, the journals in which the different members of the institutes published according to the institutional affiliation. This last criterion was used as a normalisation instrument with their respective standardisations to avoid affiliations and author names that did not correspond with the intellectual production of each member of the institute and their period of publication. The table shows that the journal that received the most publications from the institute was Acta Horticulturae with 30 published documents, followed by the Revista de Biología Tropical with 6 articles.

Table 2. Journals and publications of the institute.

Journal	Documents
Acta Horticulturae	30
Revista de Biologia Tropical	6
Check List	3
Revista Colombiana de Entomologia	3
Scientific Reports	3
Zookeys	3
Zootaxa	3
Acta Amazonica	2
Aquatic Botany	2
Biotropica	2
Brittonia	2
Diatom Research	2
Diversity and Distributions	2
European Journal of Plant Pathology	2
Genetic Resources and Crop Evolution	2
Industrial Crops and Products	2
Lwt Food Science and Technology	2
Neotropical Ichthyology	2
Wetlands	2
Acta Biologica Colombiana	1

Source: Scopus, May 2020

For the different interrelationships by research topic, a relationship was modelled for international institutions with their areas, their topics and the diversity of production according to the speciality interest. The environmental sciences revealed a sequential interaction with the biological sciences that in network analysis interact coherently with issues of biodiversity, ecology, agricultural issues, and botany, and the central themes of conservation were highlighted at this level of aggregation and in its different edges (see Figure 1).

At a more multidisciplinary and interdisciplinary level, the different related institutions prioritise research on issues of biochemistry, ecosystems, natural

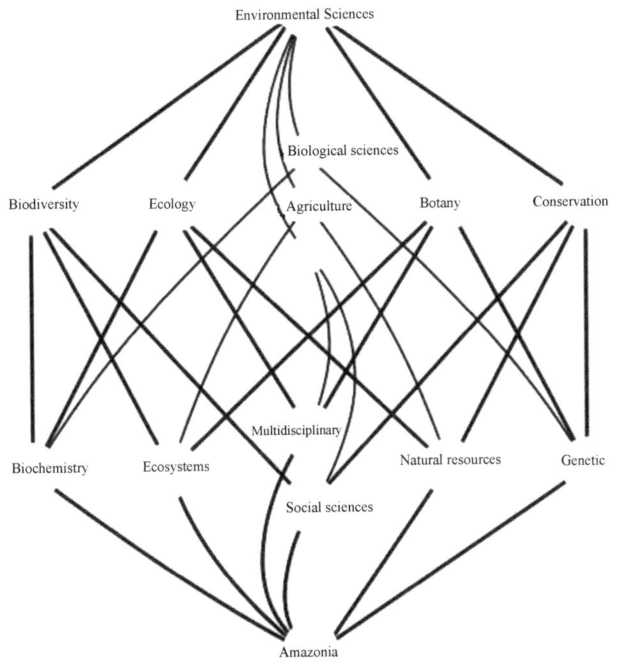

Figure 1. Interrelationships of the research topics of international institutions. Source: Own calculations from Scopus, May 2020a.

resources and advanced topics in genetics, which is treated very tangentially by different organisations at the national level. The interactions of most of the topics were adjusted by the impact on society of the various studies carried out by the institute on the Amazon in its different geographical areas of influence.

In the case of national institutions, it is possible to establish a series of interactions that, unlike international institutions, prioritise issues of a local nature, which generally have implications for the large areas investigated; for the national case, many of the investigations are established by the research topics that are financed by national organisations and by the availability of resources in each of the topics under investigation (see Figure 2).

Environmental issues showed a sequential distribution interrelationship with tropical forest issues. In research, these research areas consistently interact in networks with issues of species diversity, natural resources, topics of the current debate, such as climate change, sustainable development and ecosystems and their protection.

At the geographical level, scientific research on biogeography issues plays a predominant role in the various interactions of local institutions, biodiversity, the study of diverse populations, conservation and the incidence of deforestation. Geographical relationships play a preeminent role in studies on Amazon, especially in Colombia.

In the particular case of the institute, the multiple interactions carried out in the different indexing systems, resulting from its work on the chosen topics, which was

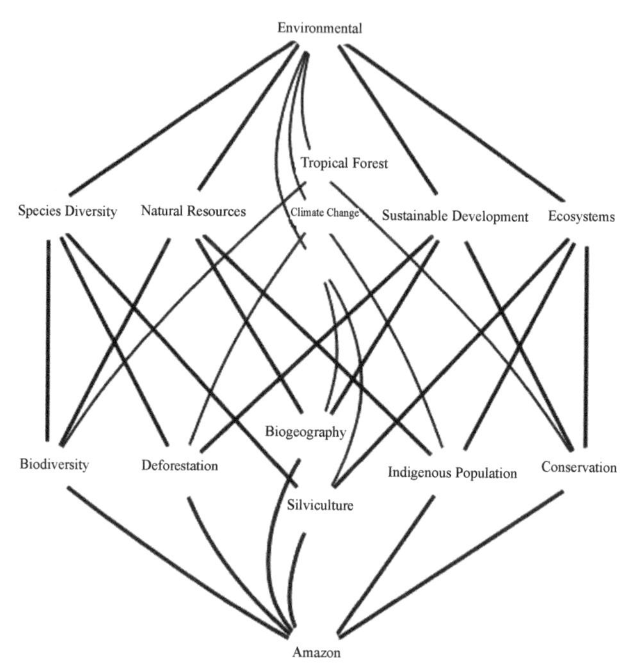

Figure 2. Interrelationships of research topics of national institutions. Source: Own calculations from Scopus, May 2020.

expressed in research projects through the visualisation and visibility of its academic production. Production is the benchmark that defines its priority areas, its activities and largely fulfils its mission.

As seen in Figure 3, the institute prioritises high-level research, which in the majority of it topics, lines, areas and projects show a relationship with international and some of the national priority issues for the coming decades. The observed interactions of the institute and the analyses carried out confirmed that the majority of its research topics were concentrated in the Colombian Amazon. This finding did not ignore that the various interactions were mainly with interest groups, and the relationships maintained by the institute revealed beneficial exchanges with the national and international scientific community. This makes the prestige of their activities a general frame of reference for their peers.

The relational themes in each of the nodes of the graph also demonstrate that there were developments that were in line with the border issues at present and that these should receive a significant degree of prioritisation and technical analysis through the implementation of quantitative methods supported by the facts and recent evidence in terms of data and mapping of behaviours and rules established in previous works by the institute, its collaborators and many of the institutions at all levels that are permanent participants in the institute.

The elaborated thematic exercise determined a series of direct connections in the relational research priorities. The nodal themes around which the institute's research revolves can be verified by their impacts on issues specific to the Amazon,

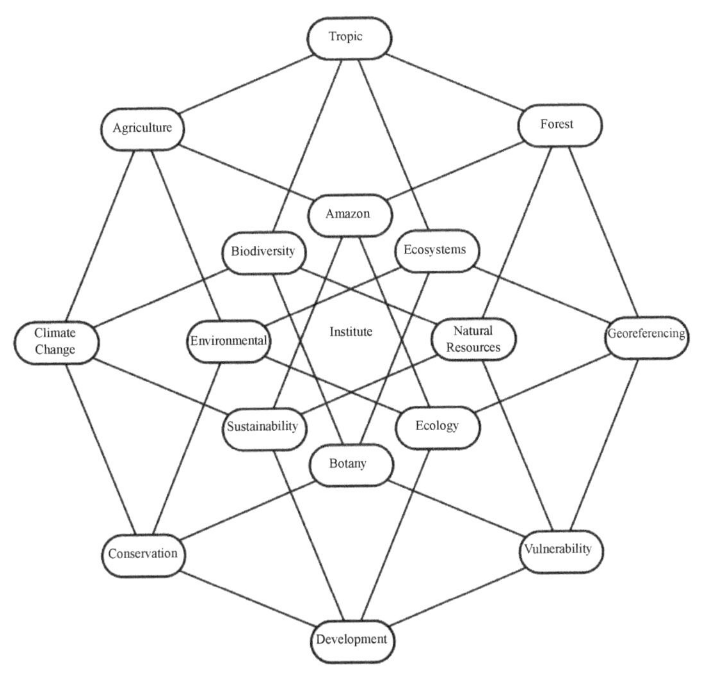

Figure 3. Interrelationships of the Institute's Research Topics. Source: Own calculations from Scopus, May 2020.

biodiversity, ecosystems, the environment, natural resources, sustainability, ecology and botany. These themes were accompanied by some that were emerging around the main themes, such as climate change, agriculture, the tropics, forests, georeferencing, vulnerability issues, development and conservation and issues associated with the population in all its aspects as well as relevant elements in regard to seeking funding in the different fields, both internationally and nationally, and sources of appropriate financial and operational stability for the institute.

Another of the topics analysed refers to the collaboration and scientific production of the institute, both internally and externally, in terms of the academic and scientific networks achieved by the institute without ignoring the indicators of cohesion in the specific area of knowledge established by the OECD. In this regard, it should be clarified that the criterion used is that of a joint collaboration between one or more institutions participating in high-level academic production of significant quality and with recognition, especially in relation to the international production.

The interactions among members of the institute were generally isolated to avoid duplication of information given the internal cohesion between members of the same area of expertise. This argument makes the analysis of cooperation between participating institutions more robust; the result of these interactions is shown in Figure 4.

From a multidimensional perspective, significant cooperation with the National University of Colombia was determined in the national collaboration; on the international side, there was significant cooperation, although there was no obvious impact on the Polytechnic University of Cartagena.

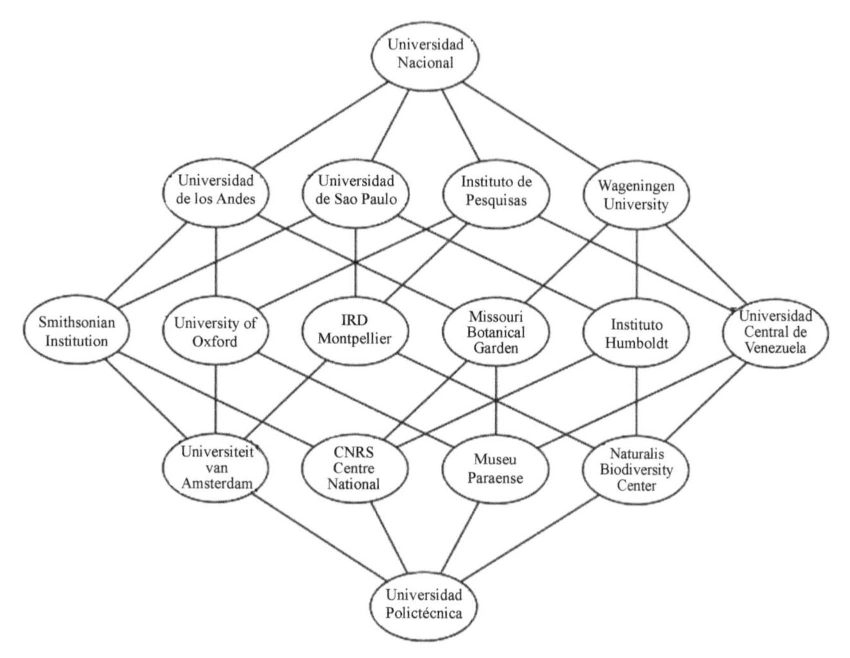

Figure 4. Interrelationships of research topics in collaboration with the institute. Source: Own calculations from Scopus, May 2020.

In an order of relative importance, the collaboration manifested itself through the Universidad de Los Andes Colombia, the Universidade de Sao Paulo (USP), the Instituto Nacional de Pesquisas Da Amazonia, the Wageningen University and Research Center and the Smithsonian Tropical Research Institute. At a lower level but without ignoring the importance of these institutions were interactions with the University of Oxford, the Central University of Venezuela, the IRD Center in Montpellier, the Alexander von Humboldt Institute for Biological Resources Research, the Missouri Botanical Garden, the Universiteit van Amsterdam and the CNRS Center National de la Recherche Scientifique and Museu Paraense Emilio Goeldi.

Prospective collaboration exists in search of consolidation with entities such as the Naturalis Biodiversity Center, the Pontificia Universidad Javeriana, the School of Nutrition and Dietetics, the Pontificia Universidad Católica del Ecuador and the Center for Ecology.

When carrying out a more refined criterion of impact and visibility of the institutions that conduct scientific studies on the Amazon worldwide, it was found that in most cases the country that conducts most of this research was Brazil followed by the United States, the United Kingdom, Germany and Holland. In Latin America, the major producers of this research were Colombia and Peru. Figure 5 determines the main institutions worldwide according to the quality of their academic production and indicates the ones that stand out the most according to production and impact indicators.

Along with the previous questions, in the case of international researchers, the authors with the highest quality and production on the subject of the Amazon were

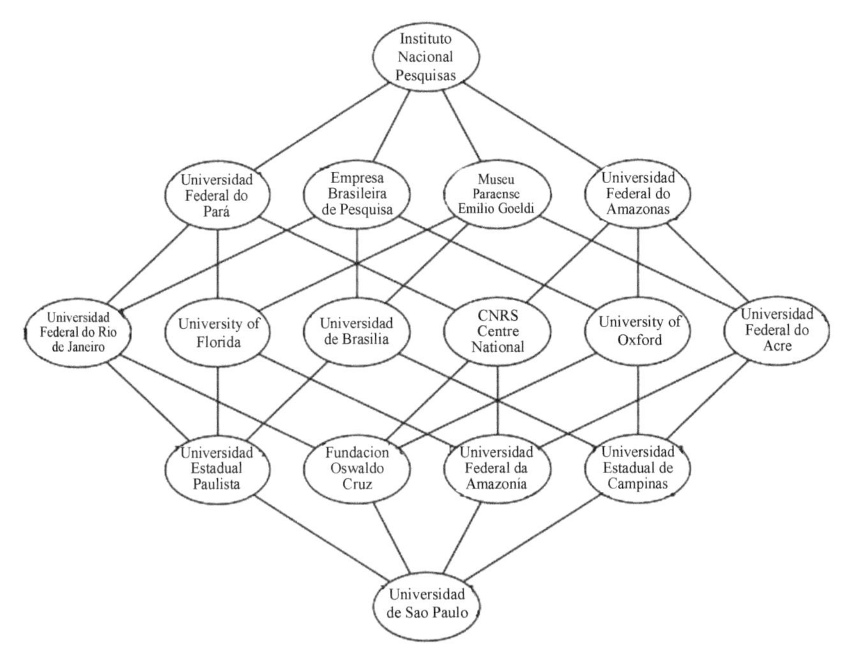

Figure 5. Interrelationships of research topics of international institutions. Source: Own calculations from Scopus, May 2020.

established. These researchers had the highest incidence on the subject based on their research and academic production and had the most referenced work according to the year of publication. Figure 6 establishes the interactions and their relationships by topic and their possible relational nodes according to the publications referred to for

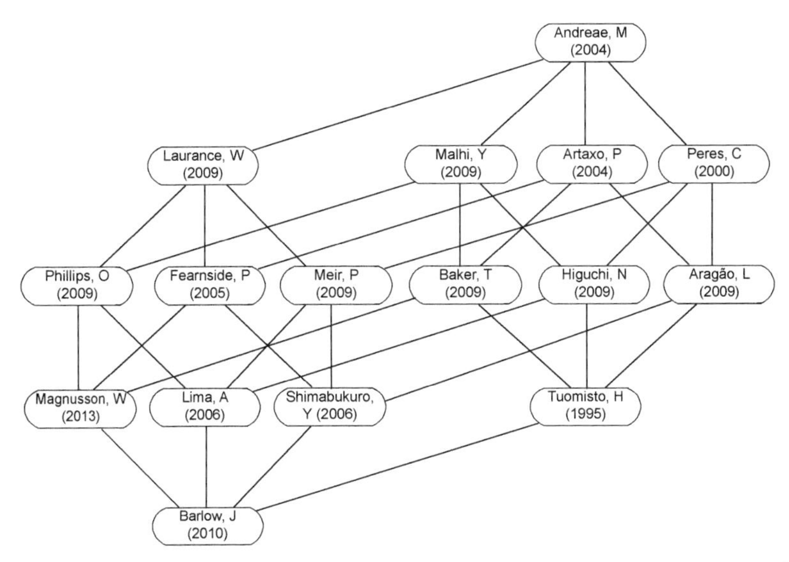

Figure 6. Interrelationships of international authors. Source: Own calculations from Scopus, May 2020.

each of the authors, which were categorised according to the publication criteria of quality and impact for their academic and scientific peers in their area of expertise.

It was noted that those of Andreae et al. (2004), Laurance et al. (2009) and Malhi et al. (2009) were the most influential studies on the subject of the Amazon, particularly with regard to forest fires and droughts that occur in that geographical space.

In a more focused situation but on issues related to deforestation, communities and forest structure, the research of Artaxo et al. (2005), Peres (2000), Phillips et al. (2009) and Fearnside (2005) stood out. Their studies were significant in terms of sensitivity to the advances of scientific research in the Amazon and provided international references on the current situation of the effects caused by the intervention in an area of vital importance for the survival of the planet and the different effects on the general well-being of the world population.

When carrying out the same exercise and using the information supplied by the institute's website with the research team as a cross-reference, the most referenced researchers of the institute were established according to the H index of citations (see Table 3). It is important to clarify that the date that the researchers were classified was May 15, 2020. It was also found that some of the researchers had multiple author IDs, which meant that there was dispersion in the authorship of the documents in the database. To aggregate the intellectual output, a single reference code was added for classification.

Table 3 defines the authors from the institute who were most referenced according to the articles with the highest citations on the subject of the Amazon. Likewise,

Table 3. Most referenced researchers of the institute.

Name	H index	Documents	Affiliation	City	Country
Researcher 1	11	18	Institute	Bogotá	Colombia
Researcher 2	6	48	Institute	Bogotá	Colombia
Researcher 3	5	18	Institute	Bogotá	Colombia
Researcher 4	5	9	Institute	Bogotá	Colombia
Researcher 5	4	10	Institute	Bogotá	Colombia
Researcher 6	3	3	Institute	Bogotá	Colombia
Researcher 7	2	7	Institute	Bogotá	Colombia
Researcher 8	2	2	Institute	Bogotá	Colombia
Researcher 9	1	3	Institute	Bogotá	Colombia
Researcher 10	1	4	Institute	Bogotá	Colombia
Researcher 11	1	2	Institute	Bogotá	Colombia
Researcher 12	1	2	Institute	Bogotá	Colombia
Researcher 13	1	1	Institute	Bogotá	Colombia
Researcher 14	1	1	Institute	Bogotá	Colombia
Researcher 15*	6	21	Institute	Bogotá	Colombia
Researcher 16*	4	5	Institute	Bogotá	Colombia

* Not reported on the Institute's website on the date of consultation.
Source: Scopus, May 2020

the table establishes the document with the greatest number of references and the year of its publication according to Scopus, which is considered one of the most rigorous and systematic databases of bibliographic references and citations, peer-reviewed literature and high-level web content. Scopus has outstanding quality with instruments for tracking, analysis and visualisation of the different investigations.

It is indicated in this table that the documents prepared by Researchers 1, 2, 3 and 4 are the investigations with the highest number of citations referring to the subject of the Amazon forest and aspects related to physicochemistry and soil biology. The works of Researchers 5, 6 and 7 were also referenced research on several of the current issues, including deforestation and its various patterns, biodiversity and ecosystems.

Table 3 determines the case of the institute the authors most referenced by the articles with the highest citations on the subject of the Amazon. Likewise, the table indicates the most-referenced document and the year of its publication according to Scopus, which is considered one of the most rigorous and systematic databases of bibliographic references and peer-reviewed literature, high impact web content and outstanding quality with instruments for tracking, analysis and visualisation of the different investigations.

It is indicated in this figure that the documents prepared by Investigator 1 (2008), Investigator 2 (2008), Investigator 3 (2008) and Investigator 4 (2002) were the investigations about the Amazon forest that were related to physicochemistry and soil biology that had the highest numbers of citations. The works of Investigator 5 (2007), Investigator 6 (2007), Investigator 7 (2018) and Investigator 8 (2006) were also studies referenced on several of the current issues, including deforestation, and the various patterns, biodiversity and ecosystems provided in these documents are national benchmarks.

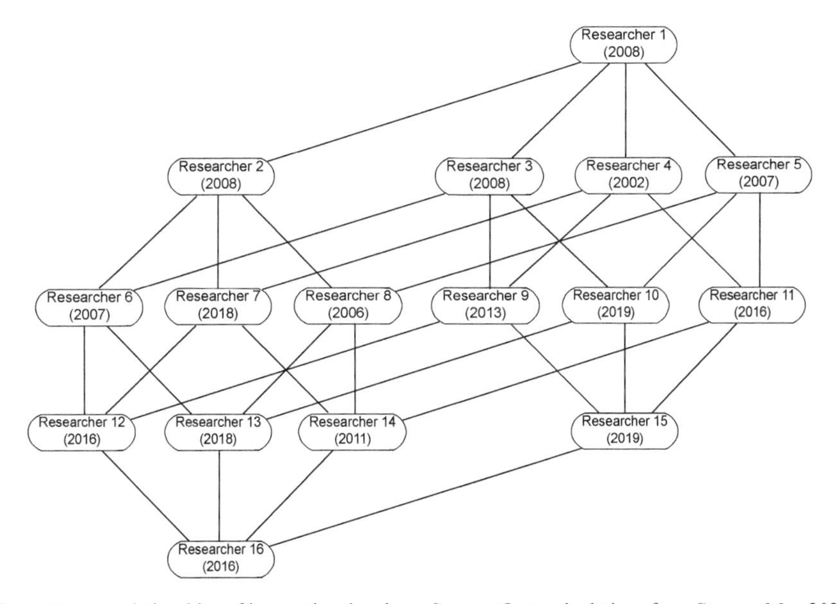

Figure 7. Interrelationships of international authors. Source: Own calculations from Scopus, May 2020.

Evolution and Trends of Research in Amazon-Related Issues and Relationships with Research Institutes

This section describes the different trends and evolution in terms of research on Amazon issues. For this, the different indicators resulting from the Scopus author identifier were used, which assigned a unique number to the groups of documents written by the same author through an algorithm that, in technical and practical terms, coincided with the authorship of most documents found in this database. For this analysis, it was applied based on certain classification criteria. Thus, for example, if a document could not be definitely matched to an author identifier, this may cause more than one entry that might exist for the same author. Aggregation by author implies the presence of duplication affecting the total number of documents, in the citational indexes and therefore in the individual and institutional authorships. For this reason, if the connection was not identified with full evidence, it was grouped separately.

Figure 8 shows the relationship between the production of the articles and the indexes of the journals in which the subject of the Amazon appeared for authors at an international level. The relationships indicate a relative concentration of a large number of articles in low- and medium-impact journals in terms of citations and references, which would generally indicate that it was possible for topics pertaining to Amazon to find better research visibility in journals at the international level with better impact indicators and referencing.

Carrying out a more detailed analysis of the relationship by means of contiguity matrices, a great dispersion was observed for the group of journals in which the subject of the Amazon was published. For the group of journals that was selected by

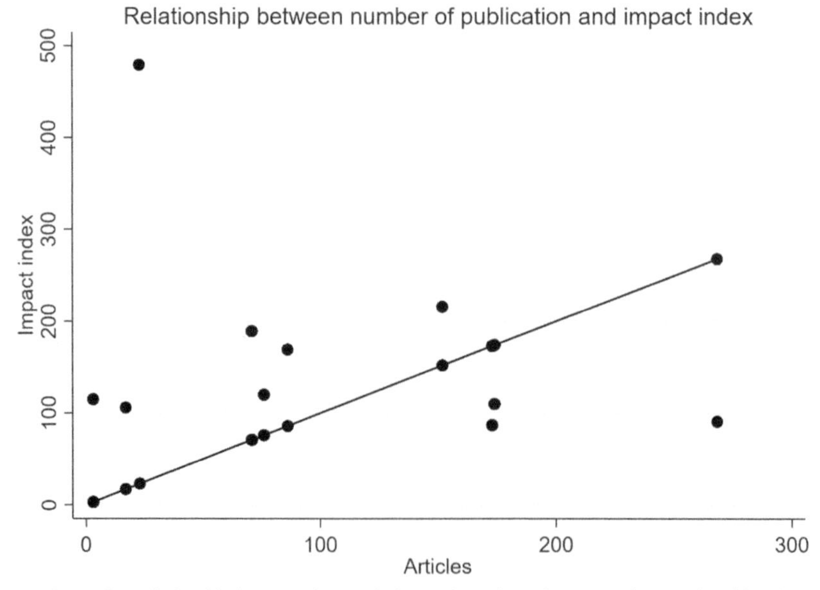

Figure 8. Relationship between impact index and number of papers at international level.

the criteria of article production and impact indicators, it was found that the journals with the highest relationship of quality to publications were Acta Amazonica, Forest Ecology and Management, Zootaxa, Biotropica, Journal of Tropical Ecology, Journal of Geophysical Research, Atmospheric Chemistry and Physics, Amazoniana and Plos One. According to Figure 9, the matrix shows that the journals with the highest impact indicators were concentrated in Plos One, Atmospheric Chemistry and Physics, Biological Conservation, Forest Ecology and Management, Biotropica, Journal of Tropical Ecology. In this sense, and given the international comparisons between the published articles, recognised authors and impact indexes, it is necessary to encourage publications with high-recognition documents that can clearly identify rigorous contributions to the area of knowledge of each of the consolidated research programmes.

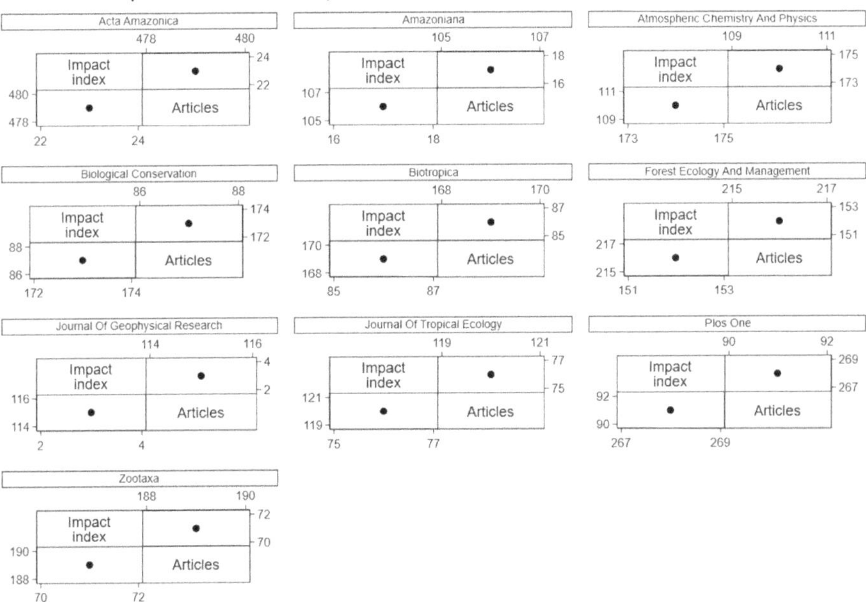

Figure 9. Relationship matrix of the impact index and the number of papers at international level.

Incorporating as one of the quality criteria the quartile in which each journal is classified, the level of each of the products can be checked according to the author(s) and institutions based on the number of articles published and the classification of the journal according to quartile. Q1 represents the highest quality classification and therefore a number of citations, Q2 represents the following classification in order of importance and quality, Q3 and Q4 indicate the medium and low-quality classifications, respectively, according to the criteria established by the main bases of data that take into account these classification indicators (see Figure 10).

As seen internationally, most of the articles related to the Amazon are published in the most recognised journals, which were classified in Q1 and Q2. This demonstrates

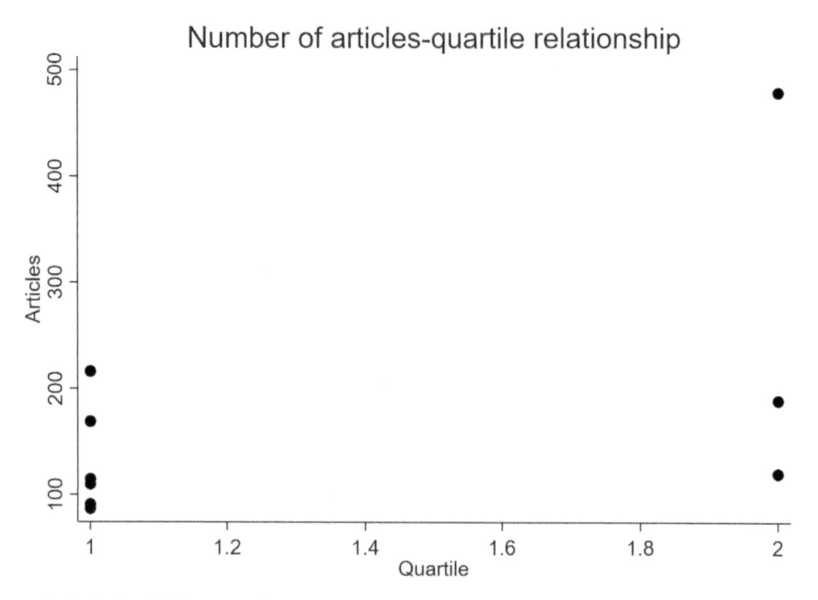

Figure 10. Relationship between the number of articles and quartile of the journal at international level.

the quality of the published studies internationally and that they are also becoming the benchmarks for the research area according to the number of citations and the citation index by each of the pairs they use as sources of information. Topics similar to their research interests were referenced in all related fields with the subject areas considered by each expert.

As a verification measure and following the same quality criteria, a series of relationships was carried out to determine the degree of concentration of academic production to establish the different weights and proportions between the production and production quality. The findings are shown in Figure 11. Approximately 50% of

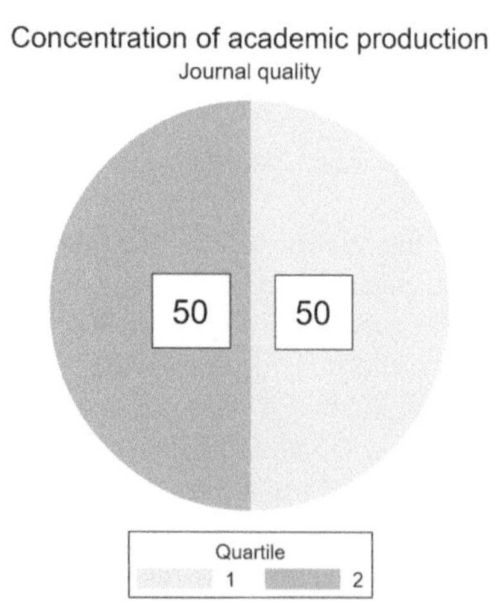

Figure 11. Relationship of production academic concentration at the international level.

the articles were published in the journals that belonged to the Q1 classification and the remaining 50% to articles were classified into Q2 according to the grouping for the comparative differential analyses.

In summary, it was found that, with the grouping criteria of visibility, citations, impact factor and a number of articles published internationally, academic production was concentrated in highly recognised journals with H indexes that ranged from 71–268. Among this classification, the journals with the most publications were Biological Conservation, Forest Ecology and Management, Biotropica, Atmospheric Chemistry and Physics, Plos One, Journal of Tropical Ecology, Zootaxa, Acta Amazónica and the Journal of Tropical Ecology.

In the case of the institute, Figure 12 shows the relationship between the number of articles published by the authors who were affiliated with an institute and the impact index of the journals in which the institute's publications appeared. The relationship showed a relative disparity between the number of articles in low- and medium-impact journals in terms of citations and references and a low number of publications in high-impact journals, which revealed that for the near future there are remarkable possibilities to improve this relationship by seeking to publish in highly recognised journals to improve the visibility indicators of an institute. This strategy would improve the quality of research and publication in magazines of greater prestige at the international level, creating an outstanding impact and benchmarking indicators.

To break the results down into greater detail and making a more pointedly separate the contiguity matrices, there was a notable dispersion in the group of journals in which the researchers of the same institute published (see Figure 13). For

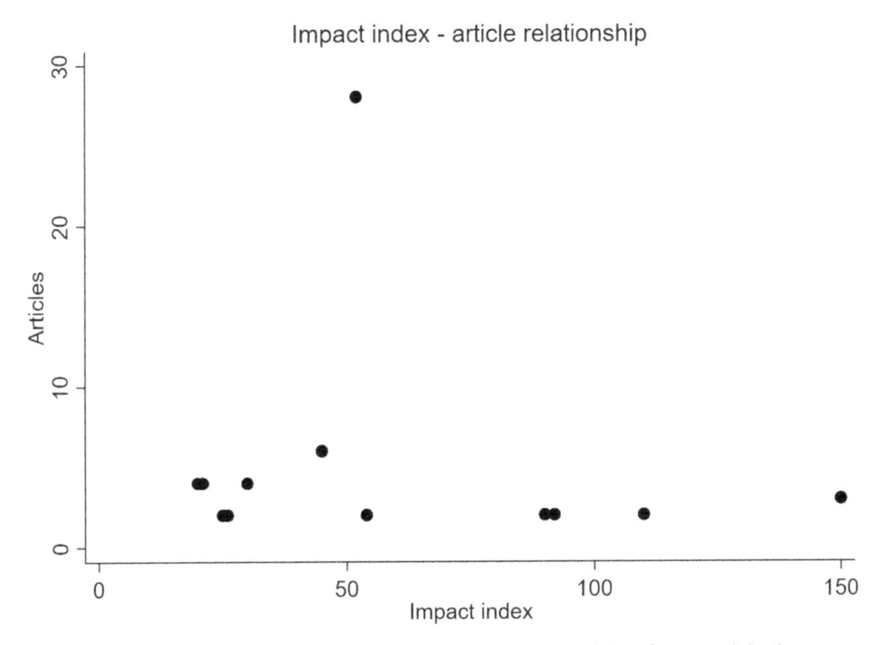

Figure 12. Relationship of the impact index and number of articles of a research institute.

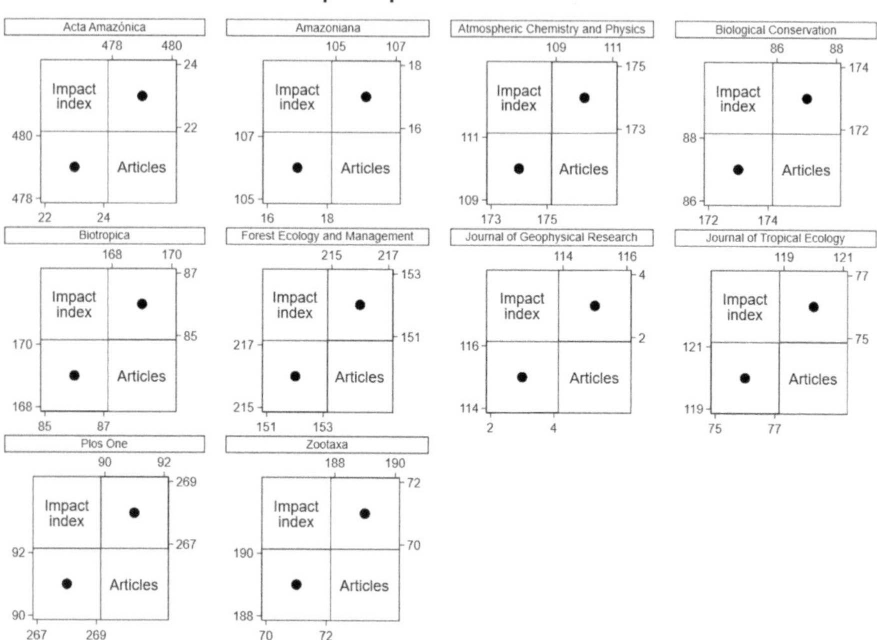

Figure 13. Relationship matrix of the impact index and number of papers of the research institute.

the selection of the different groups of journals where the research was published, it was found that there was great variability between the number of articles that were published and the impact indexes. For example, for a small significant number of articles published, the impact index of a journal could be 149, and for a significant group of publications with different themes in the same (journal) journal, its impact index could be 51. The relationship between the publication of articles and their quality was relatively low for the institute, and it was established that in aggregate terms, the journals with the most publications were Acta Horticulturae, Revista de Biologia Tropical, Scientific Reports, Check List, Revista Colombiana de Entomologia, Diversity and Distributions, Biotropica, Aquatic Botany, Genetic Resources and Crop Evolution, Diatom Research, Brittonia and Acta Amazonica.

The matrix in Figure 13 reflects that the journals with the highest impact indicator where the institute's researchers published the most were Scientific Reports, Diversity and Distributions, Biotropica, Aquatic Botany, Genetic Resources and Crop Evolution. In accordance with the above and according to international comparisons, it is necessary to encourage the production and publication of quality work in highly recognised journals to improve the institute's visibility in its areas of expertise.

Following the same criteria as in the international comparison, the quartile into which each of the journals was catalogued and established as one of the quality principles. It can be established that for each category of the magazine, the level of effort for each product to be published was greater; in other words, the requirements for the author(s) and institutions was greater according to the quartile of the journal, which was measured by its quality where the authors wanted to publish. The indicator of the journal ranking provides a measure of the impact, influence or prestige of the journals in their publication themes; in other words, it expresses the average number of citations that are weighted and that are received in the years selected by the documents published in the magazine, generally in the previous three years. For its part, the H index is the number of articles in the journal that received at least a certain number of citations during the reference period.

The relationship of the number of articles to the quartile that appears in Figure 14 shows that most of the approximately 63% of the production was published in low- and medium-impact journals, and only approximately 37% was published in journals with high impact.

From the focus of academic production aggregation, the relationship between the quality of a publication in the categorised journals can be established with the exclusion of the predatory journals. Figure 15 shows that approximately 54% of the publications were concentrated in Q4 journals, 9% in Q3, 21% in Q2 and 16% in journals categorised in Q1. As shown, it was possible to follow the trajectory of the publications carried out by international institutions and researchers and to transfer a large part of the production to high-impact journals, resulting in greater visibility for the institute.

Figure 16 shows the concentration of academic production of the institute by type of document. Approximately 78% correspond to scientific articles, 19% are conference papers, 1.8% reviews and 0.9% to notes and book chapters. In addition to the types of documents, the publications in research books in international publishers that index this production in the different databases should be explored. In the

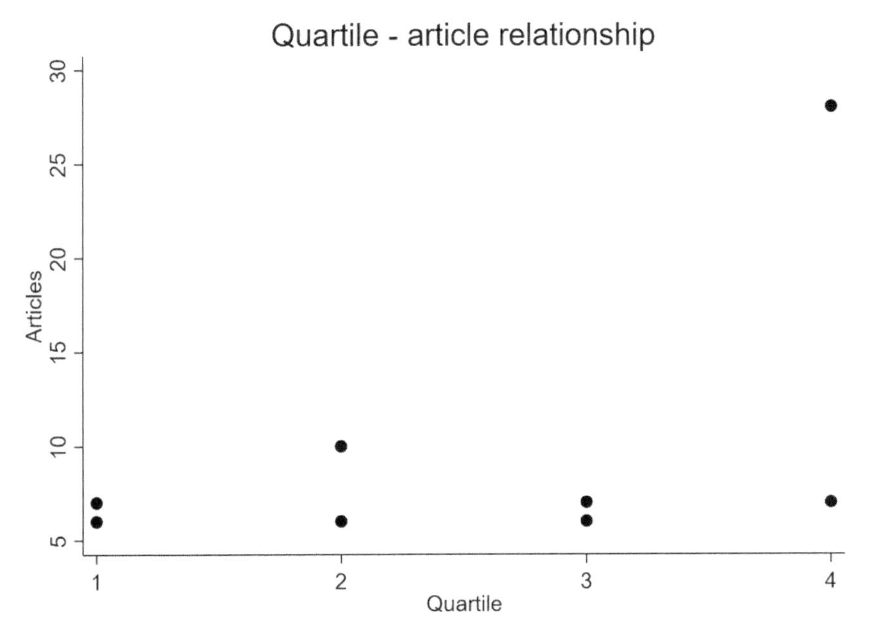

Figure 14. Relationship between the numbers of articles and the quartile of the research institute.

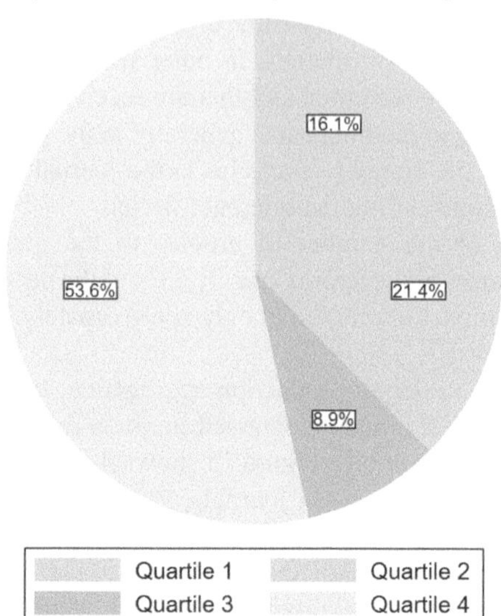

Figure 15. Relationship of the concentration of the academic production of a research institute.

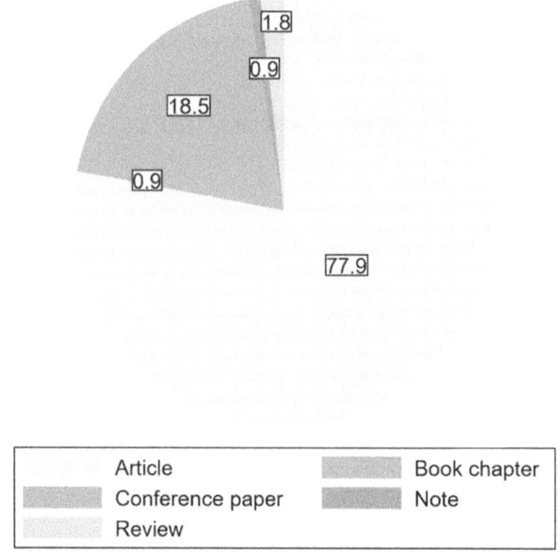

Figure 16. Concentration of academic production of the research institute by document type.

same way, the books should be accompanied by an adequate strategy for assessing publications of book chapters in international reference databases.

According to Figure 17, the sources of financing were primarily national and international. Diversification in financing is a determining factor for achieving the

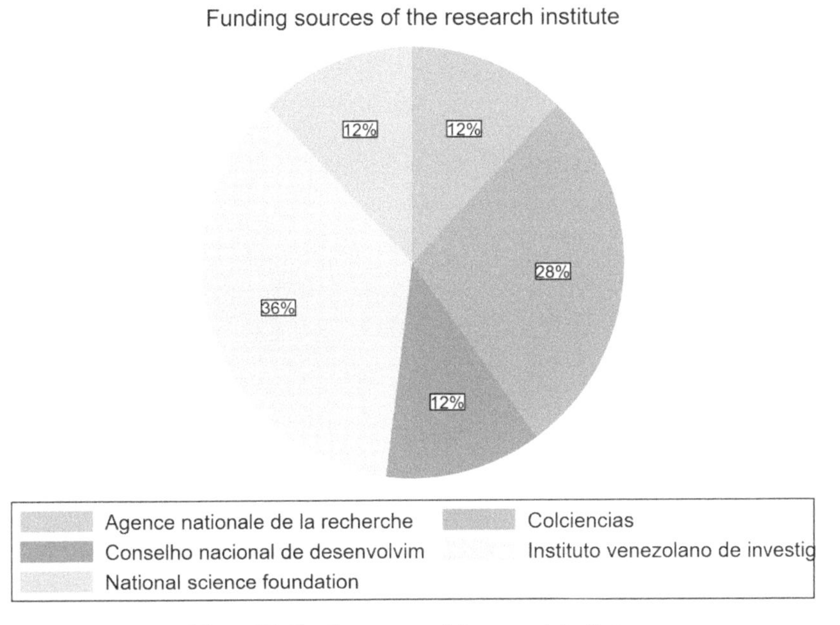

Figure 17. Funding sources of the research institute.

institute's long-term financial sustainability. Therefore, the search for sources of external and international financing for research projects should be prioritised as should working with a greater number of research networks and institutions with track records and positioning at the international level with the primary purpose of improving prestige and the reputation of the institute.

Cooperation in terms of publications is a central element for networking and contacts at all levels. One of these factors is related to the integration and advancement of knowledge. In these terms, Figure 18 showed the academic production of the Institute at the level of its most representative collaborators at the national, regional and international levels. With the Universidad Nacional de Colombia, Bogotá headquarters, approximately 23% of the academic production that was generated involved collaboration with the Universidad de Los Andes, and 8% involved other institutions around the country. Among the international institutions, Universidades de Sao Paulo and the Politécnica de Cartagena produced 5% and 12%, respectively, of the academic production.

Regarding the academic production about the Amazon by country, production was concentrated according to the degree of importance in the following most representative countries: Colombia (41%), Spain (12%), the United States (10%), Brazil (9%), the Netherlands (6%), England (5%), Peru (5%), Venezuela (5%), Ecuador (4%) and France (4%) (see Figure 19). In this regard and closely related to the sources of financing at the international level, it was feasible to obtain new resources from financing institutions with their main research institutes in these countries.

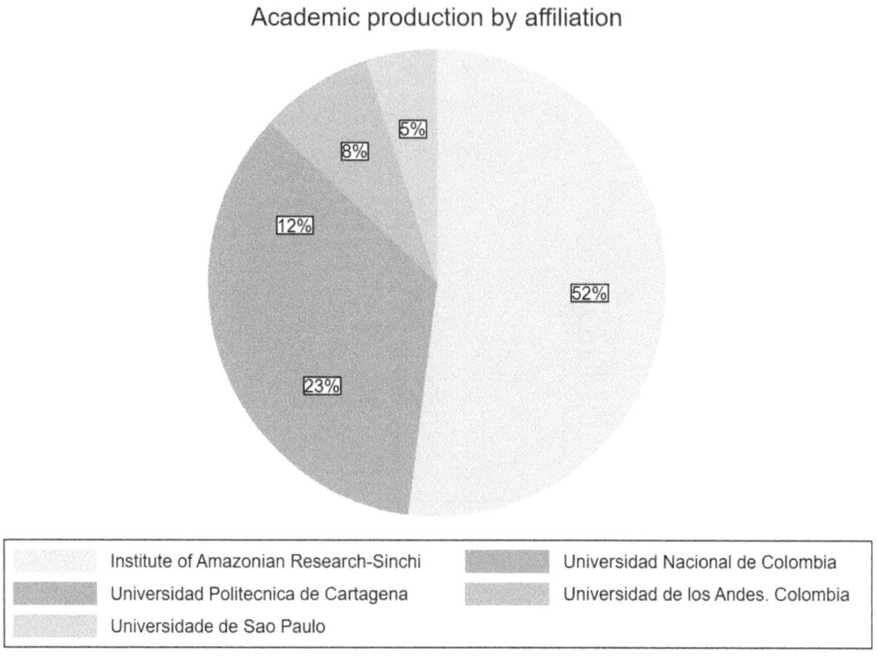

Figure 18. Academic production by institutional affiliation.

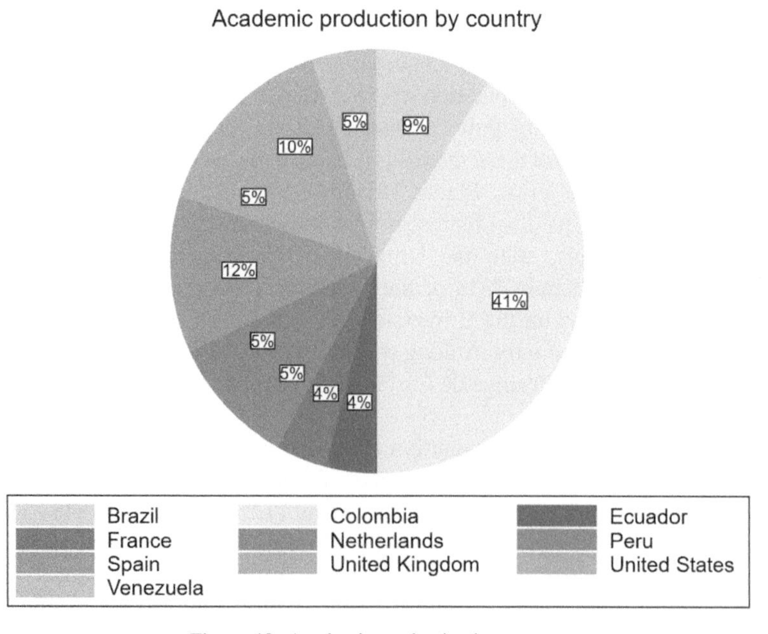

Figure 19. Academic production by country.

For example, Spain, the United States and some European countries have resources for cutting-edge research on topics, such as climate change, sustainable development, deforestation, forest management, agricultural processes, tropical forests, environmental protection, reforestation, remote sensors, biomass and cost-benefit analysis. Research associations are interested in the topics of new methodologies, conservation and protection of natural resources, sustainability, agricultural processes, forest management, deforestation and reforestation among others. Equally important is research on bioeconomic issues and new georeferencing and mapping developments at the regional level for global food security.

To support open access initiatives, electronic access to scientific publications and other series of academic materials, the Higher Council for Scientific Research created a ranking of the research centres using various selection criteria. This categorisation was prepared by the Cybermetry Laboratory of the Higher Council for Scientific Research (CSIC) based on a series of indicators linked mainly to the volume and quality of the information on its pages. It also took into account the excellent scientific production, such as the most-cited academic publications in the Scopus database and the most-cited profiles in Google Scholar.

In the most recent edition of 2019, the research institute was ranked fourth in Colombia, an important position within the most recognised research institute nationwide. Compared to the rest of the research institutes in the world ranking, the institute was in position 2,620, and Latin America was in position 124 (see Table 4).

Table 4. Ranking of researcher institutes in Colombia.

Ranking 2019	Ranking 2018	World Rank	Institute	Size	Visibility	Rich Files	Scholar
1	1	1,066	Instituto Nacional de Salud Colombia	1,425	1,627	1,645	1,729
2	2	1,267	Instituto de Investigación de Recursos Biológicos Alexander Von Humboldt	521	1,309	1,462	2,627
3	-	1,540	AGROSAVIA Corporación Colombiana de Investigación Agropecuaria	2,418	2,359	1,525	2,487
4	3	2,620	Instituto Amazónico de Investigaciones Científicas - SINCHI	3,823	3,589	2,264	2,981
5	4	2,620	Instituto Colombiano para el Desarrollo de la Ciencia y la Tecnología	120	631	2,538	3,879
6	5	2,800	Fundación Cardiovascular de Colombia	2,552	4,107	2,349	2,919
7	8	2,905	Centro Nacional de Investigaciones del Café	1,941	2,842	2,538	3,076
8	10	2,934	Instituto de Investigaciones Marinas y Costeras José Benito Vives de Andreis	1,073	1,897	2,538	3,410
9	11	3,171	Servicio Geológico Colombiano	314	1,794	2,342	3,879
10	6	3,232	Instituto Geográfico Agustín Codazzi	738	1,683	2,538	3,660
11	12	3,503	Instituto Caro y Cuervo	510	2,512	2,304	3,879
12	9	3,704	Corporacion para Investigaciones Biológicas	3,861	5,371	2,538	2,653

Source: Higher Council for Scientific Research (CSIC) - May 2020

This shows the institute's prestige, importance and recognition in all areas of its performance as a research institution with quality, reputation and influence on public policy issues at the country level. Its consolidation must be present in the future developments of its different activities to preserve its distinctive seal as a global entity.

One of the most prestigious international indicators of the different areas of the institute's work is the well-known Nature Index. This indicator generates a classification of high-quality and collaborative research results at the institutional, national and regional levels (see Table 5). The Nature Index is a database that takes into account author affiliation information that compiles research articles published in a selected group of 82 high-quality scientific journals.

Table 6 for the Colombian case classifies the main institutions that carry out high-quality research on the subjects under investigation carried out by the institute. This ranking represents an excellent positioning opportunity for the institute since it does not currently appear within any of the rankings made by this index.

The ranking of Colombian institutions in the area of Earth and Environmental Sciences is presented in Table 7. Two entities that are not higher education institutions in Colombia stand out: the Colombian Geological Service and the Alexander von

Table 5. Ranking of Colombian researcher institutes according to the nature index.

2018	Institution	FC 2018
3	Colombian Geological Service (SGC), Colombia	1.32
8	Fundación Ecotonos, Colombia	0.63
18	Instituto de Investigación de Recursos Biológicos Alexander von Humboldt	0.05
24	Centro de Investigación de la Caña de Azúcar de Colombia (Cenicaña), Colombia	0.03
26	Corporation for Biological Research (CIB), Colombia	0.01
27	Centro de Investigación Obonuco, Colombia	0.01
28	Compensation International Progress S.A. – Ciprogress Greenlife, Colombia	0.00

Source: The Nature Index – June 2019

Table 6. Ranking of Colombian institutions in the earth sciences and environmental sciences according to the nature index.

2018	Institution	FC 2018
1	Colombian Geological Service (SGC), Colombia	1.32
2	National University of Colombia (UNAL), Colombia	0.85
3	University of Valle, Colombia	0.30
4	University of Antioquia (UdeA), Colombia	0.24
5	Pontificia Universidad Javeriana, Colombia	0.20
6	Los Andes University, Colombia	0.20
7	University Corporation Lasallista, Colombia	0.03
8	Instituto de Investigación de Recursos Biológicos Alexander von Humboldt.	0.01

Source: The Nature Index - June 2019

Table 7. Ranking of Colombian institutions in the life sciences according to the nature index.

2018	Institution	FC 2018
2	Fundación Ecotonos, Colombia	0.63
9	Instituto de Investigación de Recursos Biológicos Alexander von Humboldt.	0.05
14	Centro de Investigación de la Caña de Azúcar de Colombia (Cenicaña).	0.03
15	National Cancer Institute (INC), Colombia	0.03
17	Corporation for Biological Research (CIB), Colombia	0.01
18	Centro de Investigación Obonuco, Colombia	0.01
19	Compensation International Progress S.A. - Ciprogress Greenlife, Colombia	0.00

Source: The Nature Index – June 2019

Humboldt Institute for Biological Resources Research. From this perspective, this proves to be an additional incentive for the institute to be recognized in this area of research work as an institute that makes significant contributions to the environment and to be an additional reference in these issues associated with the different areas of collaborative work.

Table 7 shows the classification of Colombian institutions that carry out research in the life sciences. The order prioritises primarily those institutions that, due to their characteristics, conduct relatively similar research to the activities carried out by the institute. In this last table, the high potential of high-impact publications focused on the topics that had verified classification criteria that were used by the index for the ranking. This would be an additional alternative to achieve better positioning of the institute in its relevant research areas. This would be even more true for institutions with very similar thematic areas of research that appear in said classification.

The trends in the production of the research institute showed a clear growing relationship over the last decades as shown in Figure 20. For the period between 1997 and 2008, an increasing trend was observed through the shift from publishing 1 article to 6 in high-impact journals. An oscillating but likewise growing trend was observed for the period between 2009 and 2018 with an extreme value at the end of the period of 26 reported publications.

The projections calculated for the 2019–2030 period were estimated and are shown in Figure 21. In the beginning, a pronounced drop was observed in the publications of the research institute to achieve relative stabilisation and a growing and significant trend in terms of academic production, particularly regarding published articles. The trend line corroborated the estimates made without major changes throughout the projected period for the prevailing trend.

In this sense, with the maintenance of the same assumptions for the different scenarios used to carry out these estimates, it was determined that despite the trend shown, there were no major changes. The smoothing growing trend was maintained without major changes in the fundamentals that generate quality production at the research institute. The assumptions were extremely conservative, but the past trend showed the publication behaviour of the institute. A less conservative scenario also showed an oscillating and increasing trend. With a higher level of productivity and academic production, for example, a change in profiles, relative improvements in

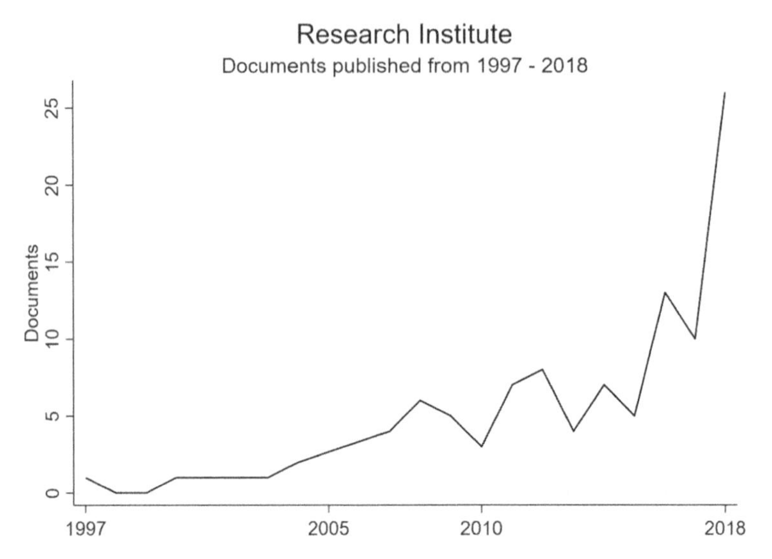

Figure 20. Papers published for the research institute, 1997–2018.

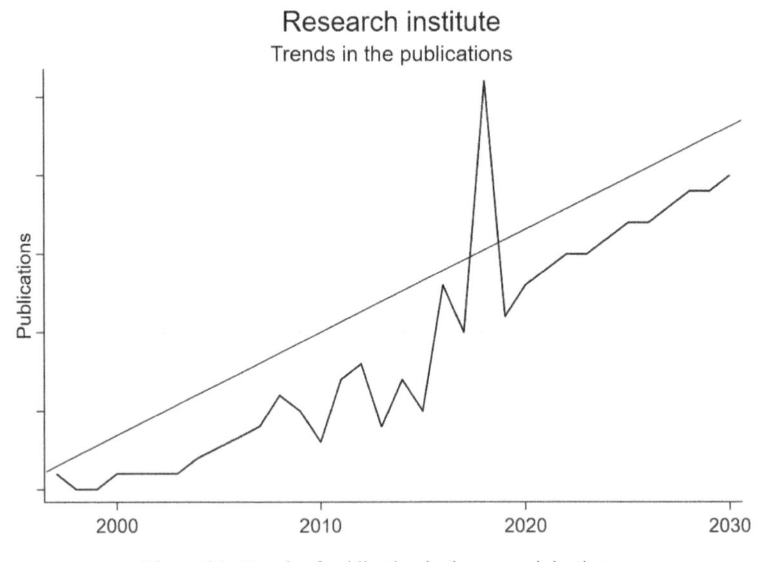

Figure 21. Trends of publication in the research institute.

human capital and gradual adjustments in the generational changeover with human capital prone to publish in high-impact journals would change both the quantity and quality of publications (see Figure 22).

At the individual level, by academic production and research, an exercise was carried out to determine individual productivity using past production. To achieve comparative data, joint production was isolated to carry out different comparisons for the respective period for each form of production and generation of knowledge, that is, individual production and in group production.

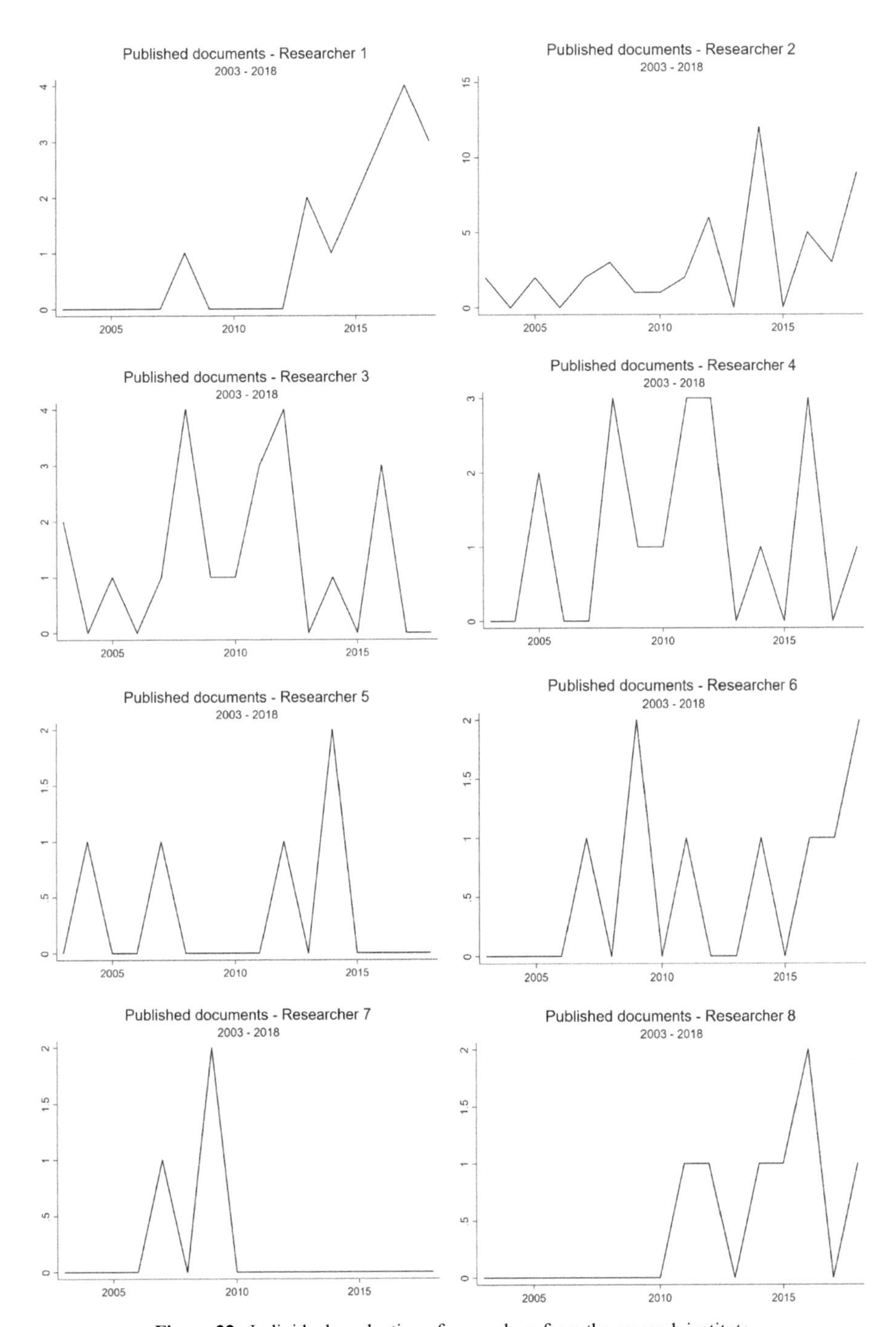

Figure 22. Individual production of researchers from the research institute.

As shown in Figure 23, there was marked variability in the academic production published by the researchers of the Institute between 2003 and 2018. The aggregate shows that for the total number of analysed observations, the average was 7.9 articles with a deviation of 6.09 and a production that oscillated between 0 and 19 articles published by all researchers in the reference period.

To increase academic production, it is necessary to create incentives to improve quality production by researchers and with it, the quantity and quality of academic production should seek diversification in high-impact and high-quality publications. It is essential to see new production possibilities academically at the book level in prestigious publishers as well as the production of book chapters in the Book Citation Index.

Figure 23 shows the different citations made to the researchers of the institute according to their publications. For the total aggregate, it was established that for the observations of the analysed period, the mean number of citations was 68.1 with a deviation of 85.47, a minimum citation value of 1 and a maximum citation value of 277 for the group of researchers selected for the period reference. As seen in the figure, the trend in citations, especially in the last decade, was increasing and with notable expectations of receiving more citations given the current situation in the Amazon, especially on the issue of deforestation.

As demonstrated for the research institute exercise, regarding the trends in publications, the individual exercise per researcher confirmed the trends indicated above (see Figure 24).

Figure 25, which shows the trends in the publications of researchers of the institute, manages to establish that in general terms with less variability in the individual trend series and despite some constant trends in the projected period, the sample showed a significant increasing trend in academic production. This again leads us to believe that incentives for high-quality production should be promoted in the near future.

According to the correlation and the projected production, Figure 25 also shows a series of increasing trends in the number of citations that the papers published by the researchers of the Institute would receive. The aggregate shows that for the total number of observations analysed, the average was 360 citations with a deviation close to 83 and a level of citations that could range from 242 to 496 citations if the number of articles published for the total of researchers in the reference period behaved in the different scenarios projected for each of the selected researchers.

In general, terms, the evidence shown, the projections made, the assumptions used and the exogenous variables analysed established the expectation of good future behaviour of the institute in terms of its academic production if a series of modifications are made to the organisational structure of scientific production and the generation of high-quality knowledge. These are necessary requirements to determine and confirm that the projections and trends behave in the desired ranges in the future based on the assumptions used.

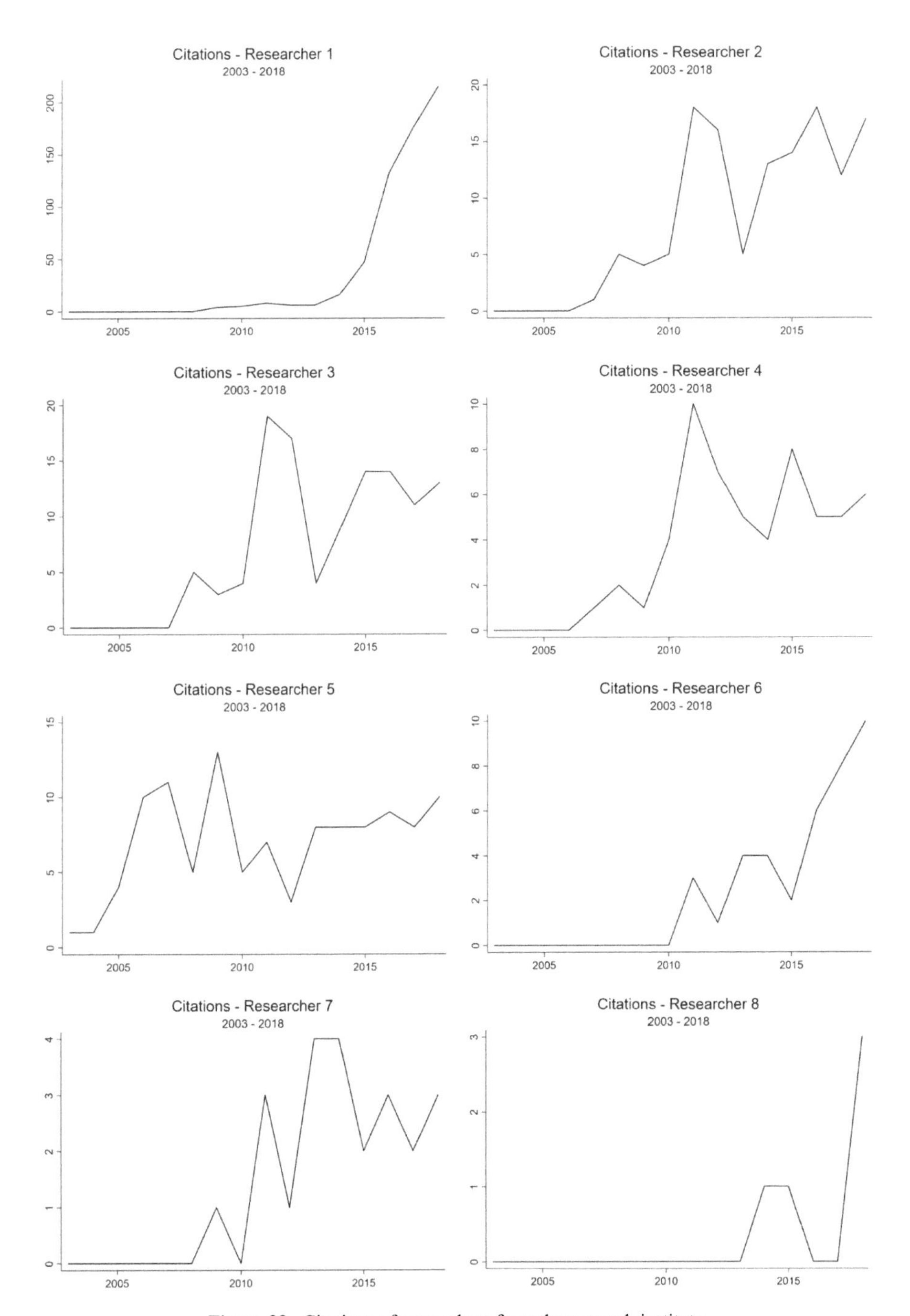

Figure 23. Citations of researchers from the research institute.

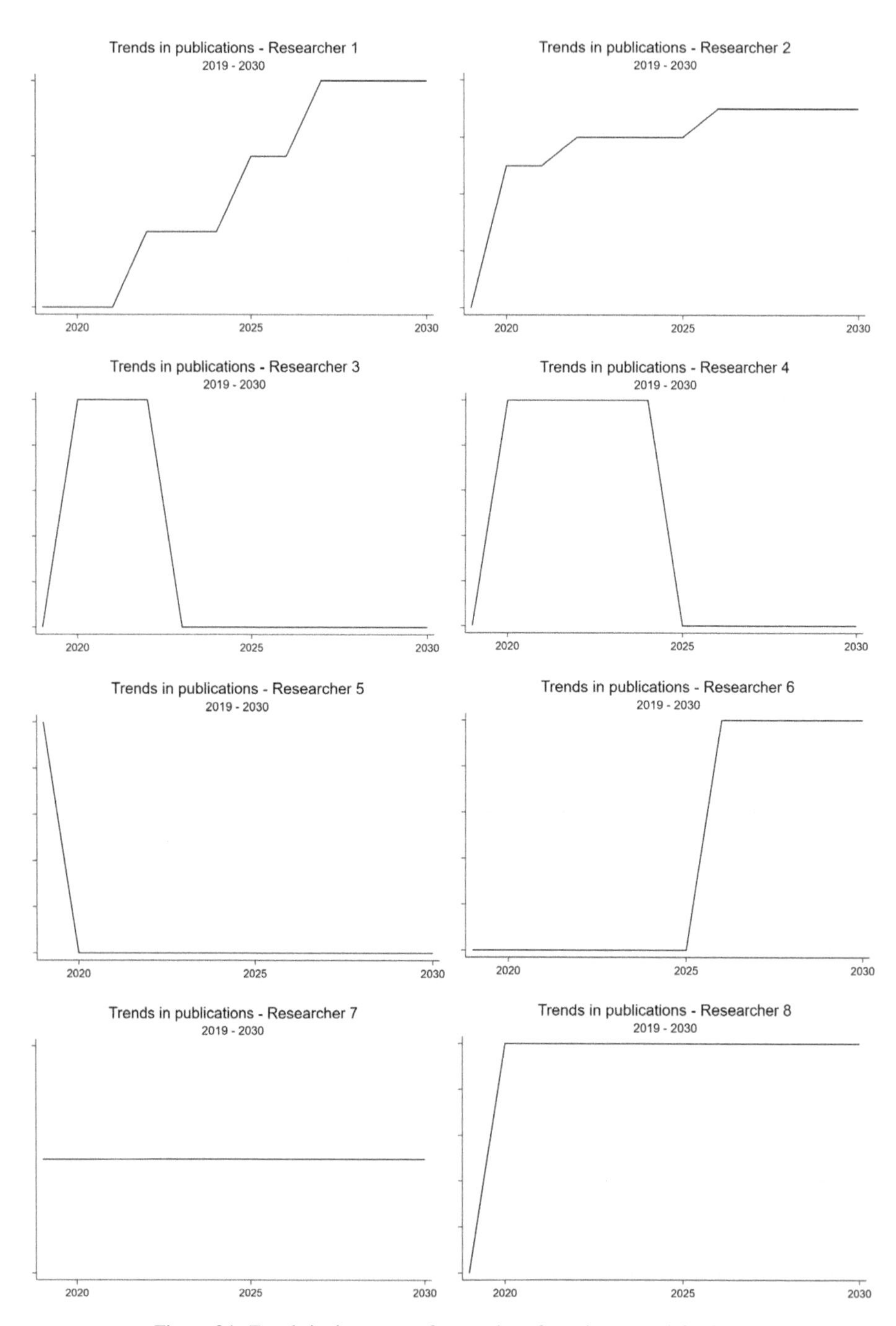

Figure 24. Trends in the papers of researchers from the research institute.

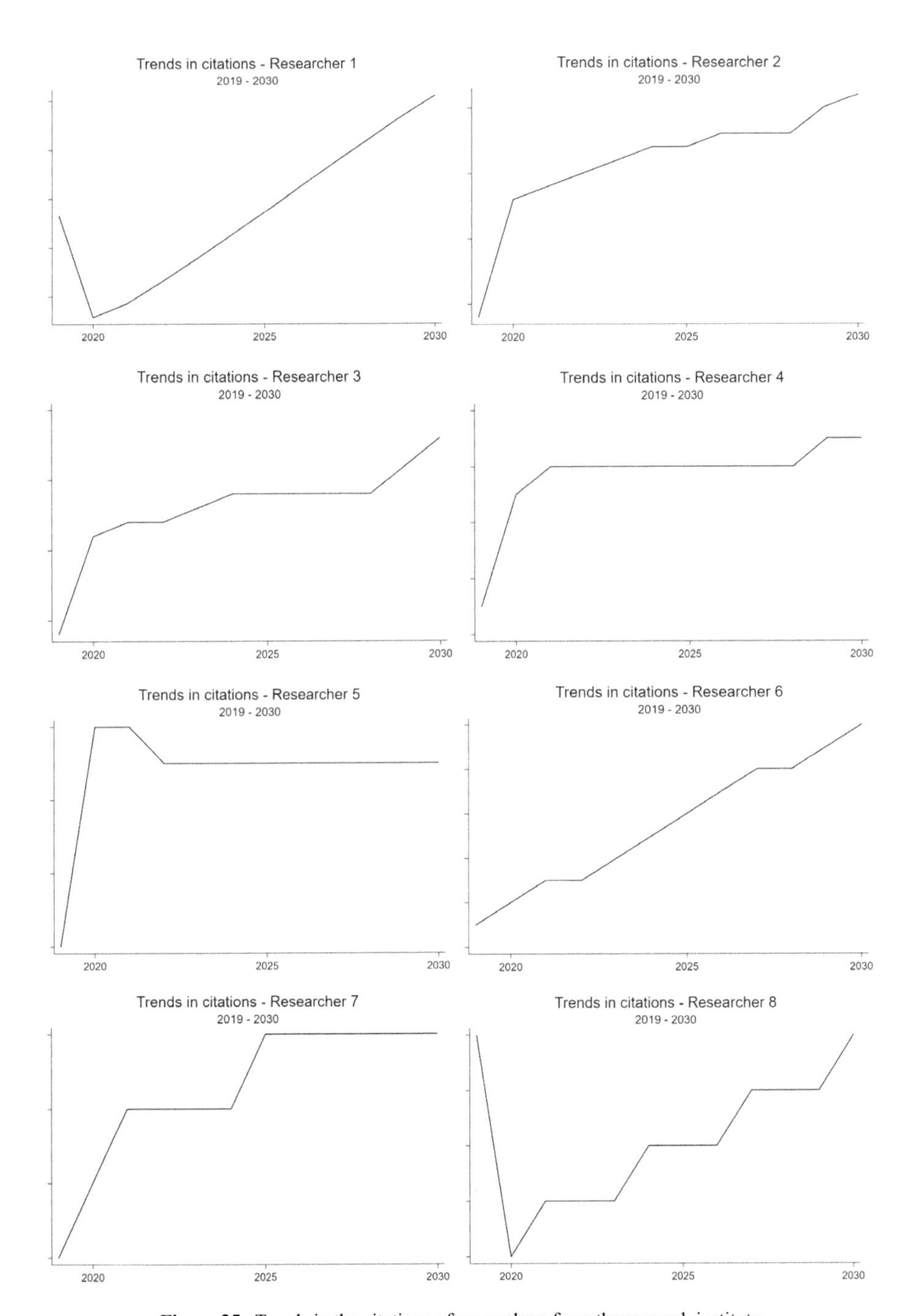

Figure 25. Trends in the citations of researchers from the research institute.

Conclusions

This analysis showed how the institute has been consolidated in recent decades as an institution that generates pertinent knowledge of its areas of expertise, achieving recognition for its research at national and international levels.

The research directions formulated by the institute and the topics the researchers work on are relevant to the global trends. However, it is essential that they be strengthened with high-quality production that guarantees the positioning and improvement of research processes as well as the diversification of their production to improve the institute's visibility in search of improved high-quality publications and better positioning in journals.

References

Andreae, M. O., Rosenfeld, D., Artaxo, P., Costa, A. A., Frank, G. P., Longo, K. M. and Silva-Dias, M. A. 2004. Smoking rain clouds over the Amazon. Science, 2004 Feb 27, 303(5662), 1337–42. doi: 10.1126/science.1092779. PMID: 14988556.

Artaxo, P., Gatti, L., Córdova, A., Longo, K., Freitas, S., Lara, L., Pauliquevis, T., ProcópioI, A. and RizzoI, L. 2005. Atmospheric chemistry in Amazonia: the forest and the biomass burning emissions controlling the composition of the Amazonian atmosphere. Acta Amazónica, 35, 185–196.

Charles, D. and Goddard, J. 1997. Higher Education and Employment—Linking Universities with their Regional Industrial Base. Paper prepared for the thematic seminar on 'Territorial Employment Pacts' in Ostersund, Sweden, 18–19 September.

ESF. 2010. Fostering Research Integrity in Europe. A report by the ESF Member Organisation Forum on Research Integrity. https://www.esf.org/fileadmin/user_upload/esf/ResearchIntegrity_Report2011.pdf.

Fearnside, P. 2005. Deforestation in Brazilian Amazonia: History, rates, and consequences. Conservation Biology, 19, 680–688.

Fritsch, M., koschatzky, K., Schatzl, L. and Sternberg, R. 1998. Regionale Innovationspotentiale und innovative Netzwerke, Raumforschung und Raumordnung, 4/98, 288–298.

Intarakumnerd, P. and Goto, A. 2016. Role of Public Research Institutes in National Innovation Systems in Industrialized Countries: The cases of Fraunhofer, NIST, CSIRO, AIST, and ITRI. RIETI Discussion Paper Series 16-E-041. https://www.rieti.go.jp/jp/publications/dp/ 16e041.pdf.

Laurance, S., Laurance, W., Nascimento, H., Andrade, A., Fearnside, P., Rebello, E. and Condit, R. 2009. Long-term variation in Amazon forest dynamics. Journal of Vegetation Science, 20, 323–333. https://doi.org/10.1111/j.1654-1103.2009.01044.x.

Malhi, Y., Aragao, L., Metcalfe, D., Paiva, R., Quesada, C., Almeida, S., Anderson, L., Brando, P., Chambers, J., Da Costa, A., Hutyra, L., Oliveira, P., Patiño, S., Pyle, E., Robertson, A. and Teixeira, L. 2009. Comprehensive assessment of carbon productivity, allocation and storage in three Amazonian forests. Global Change Biology, 15, 1255–1274. https://doi.org/10.1111/j.1365-2486.2008.01780.x.

OECD. 2014. Effectiveness of research and innovation. Management at policy and institutional levels. https://www.oecd.org/sti/Effectiveness%20of%20research%20and%20innovation%20management%20at%20policy%20and%20institutional%20levels_Meek%20and%20Olsson.pdf.

OECD. 2011. Public Research Institutions: Mapping Sector Trends, OECD Publishing. http://dx.doi.org/10.1787/9789264119505-en.

OECD. 2009. Investigating research misconduct allegations in international collaborative research projects. A practical guide. OECD Global Science Forum. http://www.oecd.org/science/inno/42770261.pdf.

Peres, C. A. 2000. Effects of subsistence hunting on vertebrate community structure in Amazonian forests. Conservation Biology, 14, 240–253.

Phillips, S., Dudík, M., Elith, J., Graham, C., Lehman, A., Leathwick, J. and Ferrier, S. 2009. Sample selection bias and presence-only distribution models: implications for background and pseudo-absence data. Ecological Applications, 19, 181–197.

Powell, P. and Dusdal, J. 2017. Science production in Germany, France, Belgium, and Luxembourg: comparing the contributions of research universities and institutes to science, technology, engineering, mathematics, and health. Minerva, 55, 413–434.

Revilla, J. 2000. The importance of public research institutes in innovative networks-empirical results from the metropolitan innovation systems Barcelona, Stockholm and Vienna. European Planning Studies, 8(4), 451–463, DOI: 10.1080/713666418.

Rosenbloom, Joshua L., Donna K. Ginter, Ted Juhl and Joseph Heppert. 2014. The Effects of Research and Development Funding on Scientific Productivity: Academic Chemistry, 1990–2009. NBER Working Paper No. 20595. Cambridge, MA.

SCImago Journal Rank. 2020. Country rankings. (database). https://www.scimagojr.com/countryrank.php.

Scopus. 2020. Detail author (database). https://www.scopus.com/authid/detail.uri?authorId=36695759200.

Stokes, D. 1997. Pasteur's Quadrant—Basic Science and Technological Innovation, Washington, D.C.: Brookings Institution Press.

Weingart, P. 2015. Die Stunde der Wahrheit? Zum Verhältnis der Wissenschaft zu Politik, Wirtschaft und Medien in der Wissensgesellschaft. Weilerswist: Velbrück Wissenschaft.

CHAPTER 3

What Can the Stock-Flow-Service Nexus Offer to Corporate Environmental Sustainability?

Luis Gabriel Carmona,[1,2] *Kai Whiting,*[3,*] *Angeles Carrasco*[4] and *Edward Simpson*[5]

Introduction

Internationally recognised management systems, such as ISO 9001 or ISO 14001, are formal mechanisms that companies frequently use to improve their corporate image and performance (Whitelaw 2012). The aim of ISO 9001 compliance is to provide products and services that fulfil consumer needs and applicable statutory and regulatory requirements (ISO 2015a). ISO 14001, meanwhile, drives improved environmental performance at the organisational level (ISO 2015b). Meeting ISO standards requires a commitment to 'continuous improvement' so that the management system becomes an integral component of long-term company operations as opposed to a 'tick box' formality. In the first few years following certification, it is relatively easy to quantifiably demonstrate progress as the company, its subcontractors, suppliers, shareholders and customers come together to identify low-hanging fruit. However, as the management system matures, it becomes increasingly difficult to meaningfully influence performance in ways that notably improve the primary

[1] MARETEC-LARSyS, Instituto Superior Técnico, Universidade de Lisboa, Avenida Rovisco Pais 1, 1049-001 Lisboa, Portugal.
[2] Faculty of Environmental Sciences, Universidad Piloto de Colombia, Carrera 9 No. 45A-44, 110231 Bogotá, Colombia.
[3] Faculty of Architecture, Architectural Engineering and Urban Planning, Université Catholique de Louvain, Place du Levant 1, Louvain-la-Neuve 1348, Belgium.
[4] Mining and Industrial Engineering School of Almadén, Universidad de Castilla–La Mancha, Plaza Manuel Meca 1, 13400 Almadén, Spain.
[5] School of Applied Sciences, Abertay University, Bell Street, DD1 1HG Dundee, Scotland.
Emails: gabrielcarmona@tecnico.ulisboa.pt; angeles.carrasco@uclm.es; e.simpson@abertay.ac.uk
* Corresponding author: kai.whiting@uclouvain.be, whitingke@yahoo.co.uk

activity (e.g., providing a public transport service) (Barrett et al. 1998, Boiral 2011, Boiral et al. 2018). Sensing diminishing returns, those leading the management system may begin to suggest changes that move attention and resources away from the main business function and focus and onto secondary or even tertiary concerns (e.g., using recycled printer paper when printing is not part of the company's core processes) (Steger et al. 2017). If left unchecked, the resulting system can become overly cumbersome and difficult to manage. It is likely to introduce unnecessary costs and complexity for minimal gain, as change increasingly becomes introduced for change sake (Gunningham and Sinclair 1999). Key resources may also get pulled from the running of the business to the running of the management system, which brings into question the purpose of being certified in the first place (Liyin et al. 2006, Weidema 2010, Reis et al. 2018).

In this chapter, we propose a stock-flow-service nexus approach as a potential solution to the plateaus experienced by the leaders of mature corporate management systems. This nexus captures the relationship between specific combinations of energy and material flows, material stocks and energy/material services (Haberl et al. 2017). Energy and material services are two interconnected concepts that consider the non-economic purpose behind the development of a business offering from a customer or societal viewpoint rather than the product or offering in its own right (Carmona et al. 2017). This perspective enables a business to consider operational effectiveness and resource efficiency using units that measure a person's ability to undertake a given activity or experience a certain state. It does not emphasise energy and materials savings *per se*, which in any case may do little to drive meaningful change. In this regard, energy and material services can be used to support an organisational level reframing of resource input and product/service output. The concept can also be employed as an integral component of a resource management strategy directed towards the Brundtland (1987) definition of sustainable development.

The move away from a traditional product or service perspective to an energy and material service perspective opens up a systems-based understanding of resource use. By considering both energy and material aspects of resource consumption and accumulation, in order to achieve non-monetary end goals, this nexus provides a more complete picture of environmental sustainability and resource efficiency (Wiedenhofer et al. 2019). This chapter thus identifies and evaluates the strengths and weaknesses of adopting a stock-flow-service nexus approach at the corporate level. The usefulness of the nexus is tested via a case study of the energy flows, material flows and stocks, and transport service provided by a bus fleet in Bogota, Colombia. The innovative aspects are (1) a demonstration of the connection between stock and flows and the service they provide to the average user of urban bus service; (2) the identification of the benefits and potential shortcomings of the stock-flow-service nexus and its associated indicators when applied to an integrated management system developed in a corporate sustainability setting; (3) a proposal of continuous improvement initiatives that could be employed by those organisations that have reached a plateau in their environmental performance.

Conceptual Framework

Energy and Material Services

Energy and material services in a general sense can, according to Fell (2017) and Whiting et al. (2020), be defined as those functions that energy and materials contribute to personal or societal activity with the purpose of obtaining or facilitating the desired end goals or states regardless of whether or not a particular flow or stock is supplied by the market. Where, the term 'function' refers to a specific characteristic which enables a person or group of people to do something (e.g., experience the mobility that transport offers) and does not refer to material properties or technical attributes such as steel's tensile strength in a chassis or a motor's RPM.

Energy and materials are not typically desired in its own right or necessarily perceived to be critical to human wellbeing (Day et al. 2016). However, because they support food production, maintain and expand material stocks, including residential buildings and road infrastructure, and provide services such as thermal comfort and mobility, it is useful to follow the energy or material production chain into services. This is especially the case if services, as an intermediate step between material production/consumption and wellbeing have the potential to improve society as a whole (Brand-Correa and Steinberger 2017, Kalt et al. 2019, Whiting et al. 2020).

The energy/material service concept allows for a distinction between material consumption or accumulation that contributes to a societal function measurable in physical units (such as passenger-km) and resource consumption or accumulation that only supports social status, financial wealth or leads to obsolete stock or waste. It is a concept that can be applied in a specific sense to several circumstances and applications. Instances include historical socioeconomic analysis, wellbeing, philosophically leaning investigations into economic activity and re-evaluation of corporate practices. In the case of the latter, end goals or desired states can be understood as a company's strategic objectives beyond financial wealth creation or economic stability.

When the energy/material services concept is used in corporate settings, it can give legitimacy to a whole range of business initiatives, including those aligned with a 'triple bottom line' perspective. The latter is a common corporate practice for evaluating performance and business value in social and environmental terms, in addition to economic ones (Bocken et al. 2014). The energy/material services concept also facilitates a strategic rethink of how energy and materials are thought of, valued and used. In particular, it breaks the paradigm of seeing materials as products and emphasises the purpose behind their manufacture and a company's raison d'être. In this regard, the energy and material service concept (and its associated tools) has the potential to support financial sustainability when the units consider aligning with a company's primary business activity. Resource savings in this area of operation will optimise and support those functions, which are integral to the corporation's continued success. This is because they are most likely to serve customers and various other stakeholders instead of subsidiary actions, which often take away resources and management focus but have a limited positive impact (and sometimes a negative one) on a company's ability to achieve its fundamental purposes.

Stock-Flow-Service Nexus

Energy and materials are two sides of the same coin (Krausmann et al. 2016). Energy production, distribution and use involve material infrastructure, while material extraction and processing require substantial energy input. In turn, a specific combination of energy flows, material flows and stock and the interaction between them creates specific types of energy and material services, including heating, lighting and transport. Adequate provision of these services then offers numerous societal benefits, both material and immaterial (Haberl et al. 2017).

Stocks, which include buildings, vehicles, machinery and electronic devices, are those materials that typically stay in the socioeconomic system for at least a full calendar year. Given their longer use phase (relative to flows, which are consumed within one calendar year), stocks fulfil many functions in the economic system (Pauliuk and Müller 2014, Weisz et al. 2015). In fact, they form the physical basis for production and consumption and are integral to the provision of social services and the generation of financial wealth (Fishman et al. 2014, Krausmann et al. 2017). For example, between 1960 and 2018, approximately 20% of Latin American economic output flows went into the construction and maintenance of material stocks (World Bank 2019). Material stocks are extremely relevant to resource accounting given that their maintenance and expansion rely heavily on diverse flow types at considerable quantities. In 2010, the material flow destined to support stock levels (52%) was greater than the annual energy, food and material flow for dissipative uses combined (Krausmann et al. 2017, Krausmann et al. 2018).

An articulated bus, like the ones featured in this case study (Volvo, B10M) requires 5.2 to 7.5 tonne of steel, around 1.5 tonne of iron, between 0.2 and 1.6 tonnes of aluminium, 400 kg of rubber and around 100 to 530 kg of plastics and other metals, such as copper and lead (Simonsen 2012). Other materials used, albeit in more minute quantities include cerium (IV) oxide for the windscreens to block the UV light and for catalytic converters. While often ignored, the sheer quantities of the different chemical elements involved mean that decisions linked to stock have considerable ramifications for corporate environmental management systems, plans, processes and overall performance.

Energy flows activate stock but do not offer service provision without material consumption (e.g., both fuel and vehicle stock combine to provide transport and later they become air emissions and solid waste). It is also important to note that the nature and quantity of both energy and material flows are heavily influenced by material stock. Fuel consumption, for example, is determined by both a user's desire to travel (the service they require) and vehicle design (the nature of the stock, including its aerodynamics, weight, etc.). In this respect, it becomes clear that there is a corporate benefit to being able to quantifiably analyse the complex relationships between energy and material flows and material stocks and how exactly they come together to provide services.

One way to analyse the physical complexity of the socioeconomic system is through the 'Stock-Flow-Service Nexus' approach (Figure 1) (Müller 2006, Haberl et al. 2017). In addition, Carmona et al. (2020) propose six nexus indicators: "stock efficiency", "flow efficiency", "stock degradation efficiency", "stock maintenance

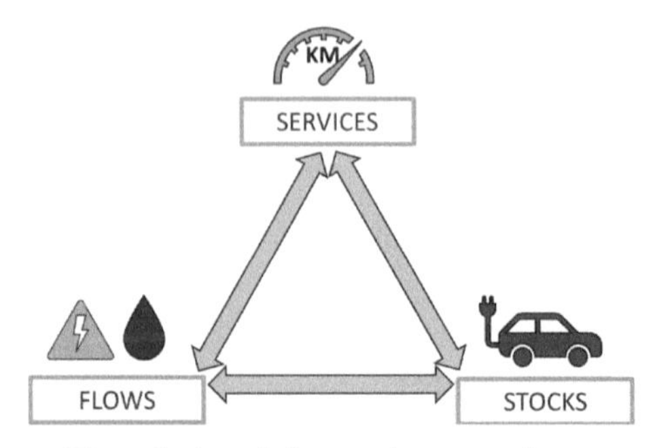

Figure 1. A stock-flow-service nexus scheme.

rate", "stock expansion rate" and "specific embodied impact". These indicators take into consideration inflow, stock and outflows from either a production or consumption perspective and should be used to quantify various interactions (stock-flow, stock-service or flow-service). Furthermore, if one only follows the trend of stock efficiency without taking into consideration the stock maintenance and stock expansion rates, one might be led to believe that service is improving at the expense of a shortage of stock when in fact this is not the case.

If a company wishes to analyse their activity from a service nexus perspective, they would need to undertake an allocation procedure in order to determine which fraction of stock/flow contributes to different services. The material service provided by some products is easier to calculate than others. For instance, smartphone use can be split up and assigned according to minutes or computer power devoted to a specific task (service). Another way to do it, especially when you have multi- and simultaneous services, is by applying allocation or weighting procedures. This something that already forms part of the LCA methodology, such as when practitioners consider environmental externalities linked to transport (Ally and Pryor 2007). The difference is that a service perspective focuses on positive outputs rather than any detrimental impacts that are not captured by more common metrics, such as GDP (as is the purpose behind extended LCAs developed by Guinee et al. 2010, Weidema 2018, Wulf et al. 2017).

A nexus approach can pinpoint exactly where an inefficiency occurs, the trade-off between stocks, flows and service and consequently the potential for overall efficiency gains. It also can provide insight into the obstacles that may be faced when trying to achieve strategic goals. The results derived from the stock-flow-service nexus in a corporate setting can thus direct/support "stock optimisation", a concept by Carmona et al. (2017) define it as the most appropriate selection of materials and their use is relative to an agreed set of criteria. The latter could be established for the achievement of environmental goals or in order to fulfil client requirements. In this respect, a company may use the nexus approach to assess current stock levels, to ascertain how stocks are being employed (or not) and whether (and to what degree) their corporate actions support societal needs.

Whilst a nexus approach can be used to evaluate the relationship between flows/ stocks and a number of energy and material services and service units, a single material service unit does not, and cannot, capture all relevant aspects of service provision. Kilometres, for instance, quantify distance travelled but are not indicative of the average user's qualitative experience. Without proper context, especially if other units are not used in conjunction, results can be misleading. A high number of kilometres is not necessarily a good thing because whilst it may mean that the transport network is large, allowing a person to travel further, it could equally signify that the network is inefficient at taking a person from one point to another, as it uses a convoluted rather than a direct route which captures more potential users but increases everyone's travel time. Larger distances will also increase CO_2 emissions and will lead to more vehicle wear and tear over a shorter timeframe. In other words, a bigger service network is not always better. In fact, one could argue that, in terms of distance, a transport service of the highest quality enables a person to travel fewer kilometres and still achieve their end goal. Thus, one should interpret service units with caution, especially in the absence of a comprehensive literature/data review and contextual analysis.

Case Study

Bogota's BRT System

This chapter explores the operation of Bogota's bus rapid transit (BRT) from 2001 to 2011. A BRT is a terrestrial mass transport system that operates on specifically designated trunk lines where all other forms of transport are excluded and ticket sale offices and stations unique to the BRT are found along the route. In this respect, a BRT with its articulated buses resembles and essentially emulates the performance and amenities of a metro or rail system but at a much lower cost (Wright and Hook 2007, Hensher and Golob 2008).

During the period studied, Bogota's BRT system was run by Transmilenio (TM), the transport management entity and various sub-contracted trunk and feeder operators, fare and user information operators and the governmental institution in charge of infrastructural development. All bus services that fell under the umbrella of TM were provided by different subcontracted companies operating within the private sector. All subcontracted activities were explicitly defined in concession contracts. All policies and practices were audited by TM.

Bogota's BRT was developed in phases (Figure 2). In Phase I (2000–2002), the system spanned 42 km across five trunk lines operated by four private companies. In Phase II (2003–2006), it spread 84 km across eight trunk lines, operated by the four original companies and three new subcontractors. These figures, following the completion of Phase III in 2012, have since increased to 112.9 km across 11 trunk lines run by 10 companies (including the seven previous ones) (Transmilenio 2017, Transmilenio 2019). Bogota's BRT was, and as of 2019, remains Colombia's most expansive form of public transport. It is the third busiest BRT in the world with 658 million annual trips (BRTData 2019). This represents a six-fold increase on the trips taken in 2001 and a 36% increase from those taken in 2011 (Transmilenio 2017, BRTData 2019). There were 2.4 million trips taken on an average weekday in 2018 (Transmilenio 2019).

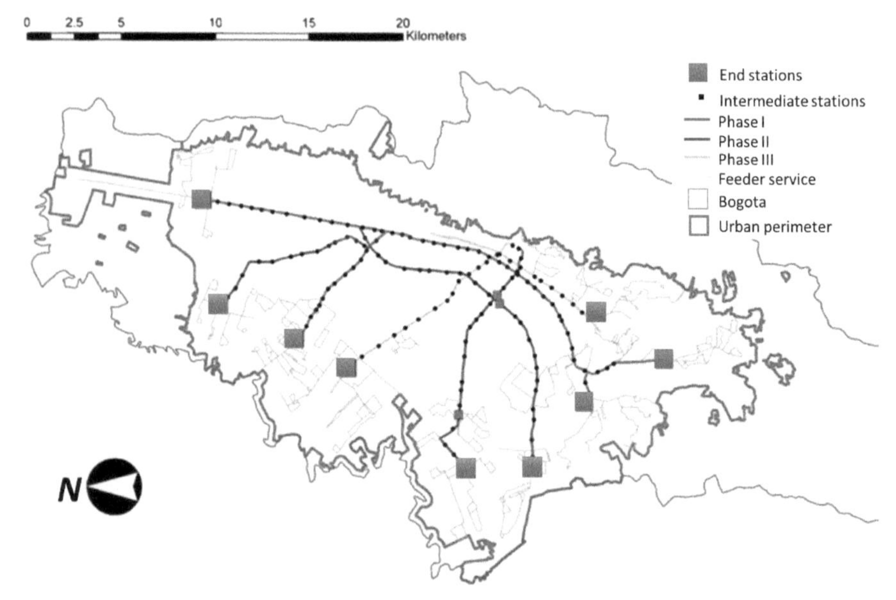

Figure 2. Transmilenio development phases. Source: Adapted from Cervero and Dai (2014).

Heightened consumer demand for BRT services has occurred due to populational increases (by more than a million when one compares the 2005 and 2016 censuses) and the conversion of centralised residential buildings into commercial ones, which has caused more people to relocate into Bogota's suburbs, thus forcing them to commute (DANE 2018, SDP 2018).

The growing demand for BRT has led to overcrowding at stations and in transit. At the same time, maintenance and operational efficiency have declined, as the replacement fleets requested in 2011 only entered into circulation in 2019 (Hidalgo 2019). Consequently, there is no business case for increasing the number of trips made by users. However, there is a case for using a nexus approach to consider efficiency from a service perspective in addition to a typical environmental analysis that considers where energy or material flow savings can be made.

Some potential stumbling blocks especially those associated with unidimensional analysis can be avoided if service is measured using a whole range of parameters and units, such as safety, comfort and punctuality in addition to distance. When measuring the latter, the nexus approach could be used by a public transport operator to highlight where the number of kilometres starts to detrimentally affect service quality. When reflecting on health and safety, one could consider the diminishing returns on the fatality rate for each additional kilogram of air emission reduction-related vehicle components. Thus, whilst we recognise the role of various metrics in evaluating service provision for the simplicity required in a proof of concept, we restrict this present case study to kilometres, as a proxy for passenger mobility. This unit was selected because aside from representing the primary function of a bus company, TM paid its subcontracted trunk line operators (between 2001 and 2011) on the basis of the number of the kilometres logged by their bus fleet, independently of how many people used the service. The onboard distance was also selected because it was measured by odometers and GPS devices, and the resulting data was independently audited by members of the United Nation's Clean Development Mechanism (CDM) scheme (UNFCCC 2015).

The Bus Company

Given the considerable size of Bogota's BRT, this case study's scope is restricted to just one of the trunk operators, which we will refer to as the 'bus company'. The bus company's fleet was constituted by 191 articulated buses of either type Euro II, Euro III or Euro IV, named in reference to European environmental norms. In addition to these vehicles, the company also owned a collection of offices and workshops where administration, maintenance and management took place. In this chapter, we focus solely on Euro II buses. These buses formed the bulk of the bus company's fleet (125 out of 191) and constitute the majority of data available on company performance between 2001 and 2011.

The bus company was managed under an integrated management system (IMS) framework. The purpose behind the management system was to develop appropriate policies and achieve the resulting corporate strategies and goals relative to the service quality parameters presented in Table 1. The company had consolidated the quality management system by 2003 and established its environmental management system by 2005. The IMS was planned under a "plan, do, verify, act" scheme with reference to international standard norms, including ISO 9001:2008 and ISO 14001:2004. The environmental aspects presented in Table 2 were formally identified by analysing process inputs and outputs, chemical use and environmental incidents.

Table 1. The parameters used by the subcontracted bus company to multi-dimensionally evaluate service provision.

Service parameters	**Variable**
Safety	Maintenance cost at the accident rate
	The percentage of programmed drivers with a maximum of four and a half hours driving shifts
	Bus replacement due to brake or driving failures
Opportunity	Fulfilment of the programmed number of kilometres
	Fulfilment of programmed bus timetable
Cleanliness	In-house cleaning fines
	Non-conformities in cleaning processes
Continuity	Breakdowns
Conformity	Fines sent by TM due to contractual non-compliance
User satisfaction	Customer satisfaction survey results
	Complaints per kilometre of service

Source: Bus company

Table 2. Significant environmental aspects during fleet operation. Source: as identified by the bus company.

Environmental aspect	**Environmental effect**	**Environmental impact**	**Situation**
Planned reserved bus distance travelled ---------------- Vehicle acceleration, braking and gear changes	Fuel consumption	Reduction in non-renewable resources	Normal
	Particulate matter and exhaust gas emission	Air pollution Visual pollution	Normal
Hazardous substance use (paint, solvents, oils, greases, diesel, coolant and aerosols) ----------- Part cleaning ----------- Resource use for the washing and cleaning of buses and associated installations	Hazardous chemical consumption	Nature resource reduction	Normal
		Quality modification of water bodies or soil	Normal
	Hazardous waste generation (coated materials, packaging and surplus of used oils, fuels, paints, coolants and thinner)	Air pollution, reduction in soil and soil quality	Normal
Substance transfer (paints, solvents, cleaning supplies, coolant, oils and fuels) ------------------- The breakdown of vehicle systems due to road-based impacts with objects or crashes ----------- Electronic system failure ------------ Terrorist acts	Risk of spills/leaks	Water pollution and soil contamination. Reduction in water body quality	Abnormal
	Risk of fire and/or explosion	Air pollution or soil contamination	Abnormal/ Emergency

Source: As identified by the bus company

In 2007, the bus company established, as part of its IMS, a set of policies and practices that could be considered as a rudimentary example of nexus thinking from a service perspective. This approach was adopted when the bus company realised that they were mismanaging fleet mileage. The rationale behind this action was to ensure effective and standardised use of the bus fleet by guaranteeing that each bus met the projected distance travelled (as contractually stipulated). The business case was the prevention of unnecessary vehicle scrapping, reduced maintenance time and the lower costs associated with further procurement of spare parts and potentially individual buses. The company's goals were achieved via a stock optimisation model (based on lineal programming) that projected various scrapping scenarios relative to expected daily kilometres, running costs and maintenance activities for each bus. The main indicator used to measure the success of the intervention was 'Attainment of Projected Distance for the Selected Scenario'. The result was that no bus was scrapped due to travelling beyond legal limits, although there were some unexpected consequences including increased fuel costs.

Method

In this case study, the stock-flow-service nexus approach is used to assess the degree at which flows and stocks were influenced by fleet mileage in the business as usual scenario (2001–2007) and upon enacting the stock optimisation strategy (2008–2011). Five indicators are employed to verify the added value in approaching sustainability issues from a service nexus perspective.

Quantifying Energy/Material Flows, Stocks and Services

The data presented in this case study is obtained from the following primary sources: TM and the subcontracted bus fleet operator. The raw data, which correspond to empirical processes, were internationally verified and validated by the United Nations. The obtained results represent a snapshot of company operations and not a theoretical model based on estimations and/or assumptions. In this respect, it is important to note that these results are specific to the operating conditions (selection of vehicle, percentage of biofuel mix, maintenance frequency, the origin/destination matrix, etc.) of a particular bus company during the period studied.

The primary input, at 97 to 99% of the total energy consumption, was diesel fuel. This is normal given that it represents 74% of the energy units employed in the mobilisation of passengers travelling on Latin American public transport (Carmona and Ocampo 2014). What is atypical is the fact that biofuel was added to the diesel, when Colombia passed the resolution 182142 of 2007, which enforced a 5% addition of biofuel into the conventional fuels used by diesel motors. This legislation established the technical quality of the fuel's specifications. It also led to the development of B5 fuel, a name chosen in reference to the percentage needed. In 2012, this legislation was superseded by one that required a 7% biofuel addition to diesel, the implications of which lie beyond the scope of this chapter. The bus company's fuel consumption was registered daily at service stations via the i-button system, which served as inventory control. This real-time system registered and stored information on fuel supply type and quantity and distance travelled, amongst others. This data was audited under the United Nation's CDM project.

Non-fuel materials inflows (e.g., lubricants, greases and coolants) were not considered within the scope of the study, since the fleet in question did not increase, and such flows only become relevant when stock expands (or contracts). Whilst higher volumes of material inputs were used during the maintenance phase, they did not directly impact service provision and were adequately captured in the outputs in the form of waste. The latter are categorised as 'conventional waste' and 'hazardous waste'. Waste streams include oil, air filters, oil filters, contaminated material, batteries, mud, used thinner, scrap metal, coolant, fluorescent tubes, used tyres, rubber, broken/replaced glass, acrylic, electronic waste, plastics, aluminium, polystyrene and PET.

Stocks refer solely to Euro II buses and do not encompass other vehicle types, buildings or other forms of infrastructure or machinery that supported the bus company's operations. Mobility as a service was measured in kilometres and captured in a legally fitted tachograph, a device that records distance covered, vehicle speed, vehicle operation time, driving time, work disruptions and rest periods.

Stock-Flow-Service Efficiency Indicators

Table 3 summarises the indicators used in this BRT case study to express various aspects of the stock-flow-service nexus. The 'stock efficiency' indicator (Equation 1) identifies the impact of stock size on service provision. It can be used to demonstrate the significance of material accumulation and consequently may drive corporate policy, practice and procurement decisions. It can also identify where service units need to be adjusted so to maintain or improve service efficiency. In a transport case study, this indicator provides insights into how vehicle size and design influence service delivery. For instance, it can reveal whether the current bus fleet is large enough (or not) relative to the number of kilometres of service requested by the public transport authority. The 'flow efficiency' indicator (Equation 3) highlights how product innovation and/or user interaction affect resource consumption and service provision. It can be used to redraw public transport policies linked to sustainable fuel transitions because it is when considering stock efficiency relative to flow efficiency, that the trade-off between flows and stocks becomes apparent. The 'stock degradation efficiency' indicator (Equation 5) states the level of physical depreciation that occurs during service delivery. It can help identify whether the waste is being generated due to preventive maintenance activities or due to poor road conditions or driver error.

The 'stock maintenance rate' (Equation 7) depicts the minimum amount of material flow that is required to maintain/upgrade stock (rather than expand it). It provides insights into stock component longevity and thus helps decision makers

Table 3. Stock-flow-service indicators.

Indicator	Description	General Equations	Case Study Application
Stock efficiency	The amount of stock required to provide a unit of service	$\frac{S}{M_{Stock}}$ (1)	$\frac{Service\,(km/year)}{Stock\,(number\ of\ buses)}$ (2)
Stock degradation efficiency	The amount of stock that degrades (worn out/made obsolete) to provide a unit of service	$\frac{S}{M_{Out\,flow}}$ (3)	$\frac{Service\,(km)}{Waste\ out\ flow\,(tonne)}$ (4)
Flow efficiency	The amount of inflow that is directly consumed to provide a unit of service	$\frac{S}{M_{In\,flow(consumables)}}$ (5)	$\frac{Service\,(km)}{Fuel\,(kg)}$ (6)
Stock maintenance rate	Fraction of material required to maintain stock at a specified level	$\frac{M_{Out\,flow}}{M_{Stock}}$ (7)	$\frac{Steel\ out\ flow\,(tonne/year)}{Steel\ stock\,(number\ of\ buses)}$ (8)
Interquartile range (IQR) of stock evolution	Distance between first and third quartile of the different elements that constitute the stock	$Q3 - Q1$ (9)	$Q3$ *kilometre distance* − $Q1$ *kilometre distance* (10)

Note: Where, S: Material service, M_{Stock}: Material stock, M_{Inflow}: Annual material inflow (depending on the equation it may be consumable or durable), $M_{Outflow}$: Annual material outflow, Qn: quartile n to anticipate future procurement, for instance, for new buses or spare parts. While out of the scope of this chapter, an additional resource interaction can be captured through the 'specific embodied impact' indicator. This identifies the number of inputs or outputs associated with a flow or stock. One can use it to determine how much CO_2 is released from fuel consumption, stock production and maintenance.

Finally, the 'interquartile range of stock evolution' (IQR – Equation 9) represents the statistical dispersion (the spread of the middle half of the data) between the first and third quartiles relative to specific stock characteristics, such as year of registration, kilometres, etc. It can identify the intensity or frequency at which an individual bus provides a service. In this case study ensuring that all buses were employed to the same degree prevented vehicle scrapping and unnecessary reinvestment in new buses, which represented a considerable financial and environmental cost.

Results and Analysis

Flow, Stock and Service Trends

Figure 3A shows that the service provided by the Euro II fleet varied between 7 and 11 million km/year with the maximum annual distance covered by the bus company's fleet in 2004. This peak occurred as the BRT network expanded prior to newly contracted operators being able to fully adapt to their new role. A 5% decline between 2004 and 2006 followed as the new operators began running at full capacity. This new dynamic gave TM the opportunity to redesign bus routes and reassign them across all operators, according to their fleet capacity. In 2007, the bus company purchased newer vehicles (Euro III and IV) after an amendment to

Figure 3. Service, flows and stock variables from 2001–2011. A: Service in km. B: Inflow as tonnes of fuel consumption. C: Stock as bus units. D: Outflows as tonnes of wastes.

the service contract. This resulted in a reduction in route assignments for the older buses within the fleet. In 2011, for instance, the Euro II vehicles were responsible for only 60% of the total kilometres travelled. As one might expect, Figure 3B shows a strong correlation between service and fuel consumption and thus the most intense diesel use (5,758 tonnes) occurred in 2004. At the same time, Euro II bus stock (Figure 3C) remained constant with only one bus being scrapped due to a traffic incident. Figure 3D presents an absolute increase in waste production because with an ageing stock (and even if the number of vehicles remains constant and the service reduces) there is an increasing amount of energy and material flow associated with maintenance activities.

Trends in Nexus Indicators

Relative to 2006 (2.046 km/kg), Figure 4A presents a 10.5% fuel efficiency increase from the 1.851 km/kg baseline registered in 2001. This incremental improvement resulted from improved fuel injection procedures and control measures designed to keep the percentage of kilometres driven by reserved service buses (those running from point A to B but not operating a public service) to less than 2% of the total. From 2007 onwards, fuel efficiency reduced to 1.948 km/kg by 2011. This happened for two reasons. The main one was the introduction of B5 diesel, which was of poor quality due to the unexpected presence of solids that were not removed in the production process. These solids negatively affected the injector system and reduced combustion engine efficiency. It also had detrimental environmental effects because it increased the opacity of exhaust gases (Carmona and Ocampo 2014). The second

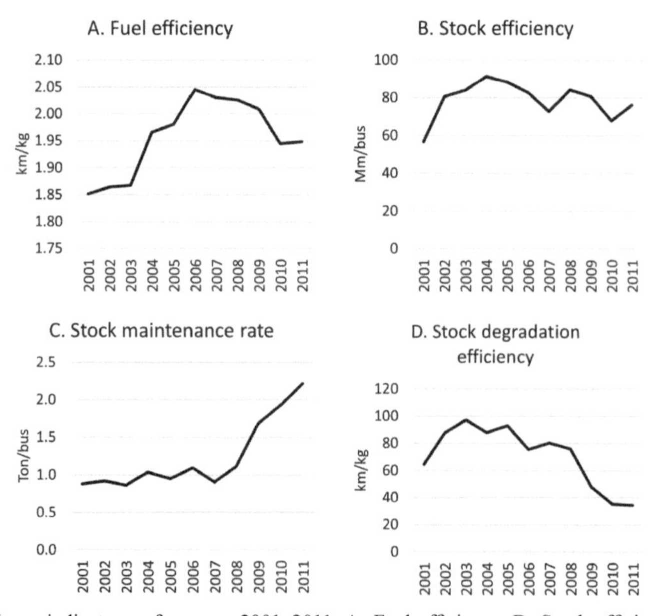

Figure 4. Nexus indicator performance 2001–2011. A: Fuel efficiency. B: Stock efficiency. C: Stock maintenance rate. D: Stock degradation efficiency.

reason was an inability to maintain reserved service bus kilometres at 2% following TM's route redesigns. By 2011, such kilometres had reached 7% of the total.

Figure 4B shows the yearly distance travelled by the average bus. In 2001, the stock efficiency was 57,000 kilometres per bus. This number increased to 76,000 kilometres per bus by 2011. Given that stock remained almost the same, any fluctuations occurred due to changes in service provision, in the sense that stock efficiency increased in direct proportion to kilometres travelled. The same held for the stock maintenance rate, which was only affected by the quantity of waste generated (Figure 4C). For the latter, and taking into consideration that each bus weighed 9 tonnes, the weight equivalent of 10% of the fleet stock was required in 2001 to maintain/support bus operations. This increased to 25% by 2011.

In summary, the company's action plans regarding improved operational efficiency were countered and masked by various external factors including TM's expansion of trunk lines and route redesign plus the legally enforced use of poor quality biodiesel. Internal factors, particularly those linked to increased vehicle deterioration due to ageing, also contributed to the problem. Consequently, although improvements were made, they simply prevented overall performance from worsening.

Stock degradation efficiency went from 65 km/kg in 2001 to 34 km/kg by 2011 (Figure 4D). Figure 5 presents the stock degradation efficiency for each waste stream. Waste quantities of used oil, which was replaced every 7,500 to 10,000 km, is coupled to service provision; meaning that the more a bus travelled, the more waste oil was produced. Other waste streams were influenced by maintenance cycles (e.g., engine and injection system repairs, battery and tyre replacements), hence the peaks. Waste was also produced in the corrective maintenance activities that followed accidents. Outflows derived from the latter were constituted by contaminated materials, thinner and cardboard and produced an irregular peaked pattern. It is important to note that only one bus suffered irreparable damage following an accident in 2003. This event was removed from the data as it was an outlier that produced 18 tonnes of additional outflow. The quantity of waste per bus increased as the fleet aged due to a higher demand for vehicle maintenance. In 2011, 36 waste streams were generated. The most common ones were used oil (27%), scrap metal (17%), used tyres (9%) and mud (9%). Some 56% of the waste's total weight was sent for recycling.

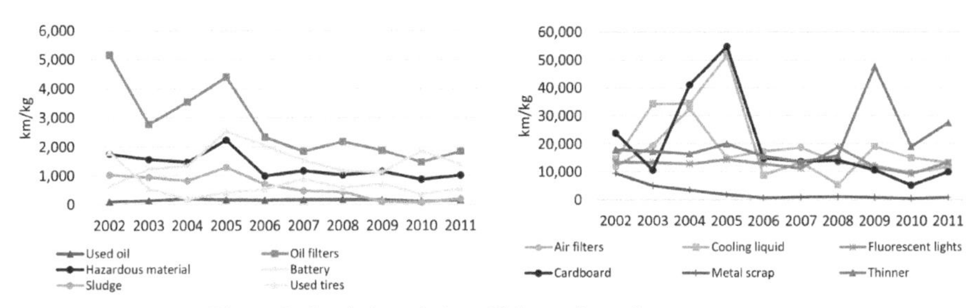

Figure 5. Stock degradation efficiency for each waste stream.

Stock Optimisation

Figure 6 presents the results of the stock optimisation scenario as measured by the IQR. The optimal scenario is a low IQR because it means that the distance travelled among the fleet was uniform. The high IQR values registered at the beginning of BRT operations followed by their rapid decline and stabilisation were an expected part of the learning curve. Between 2001 and 2006, the distance travelled by any one particular bus was not monitored relative to others in the fleet. Consequently, some buses were used at a much higher frequency than others. Those with the least mileage were driven only during peak hours whilst those used the most were seldom parked. If this practice had continued, some buses would have reached the legal limit of a million kilometres and would have been scrapped. This would have rendered the company unable to provide the agreed-upon level of service during the peak period without investment in new vehicles. In response to this imbalance, an optimisation strategy was launched in 2007. The impact of the strategy can be seen in the IQR's decline between 2009 and 2011; however, it did not automatically transform into better results in terms of fuel efficiency, stock efficiency or stock degradation efficiency (as seen in Figure 4). This highlights one of the challenges of improving environmental performance when, as is often the case in an integrated management system, it does not rely on a single action but instead encompasses various complex interactions (not all of which are captured by the stock-flow-service nexus).

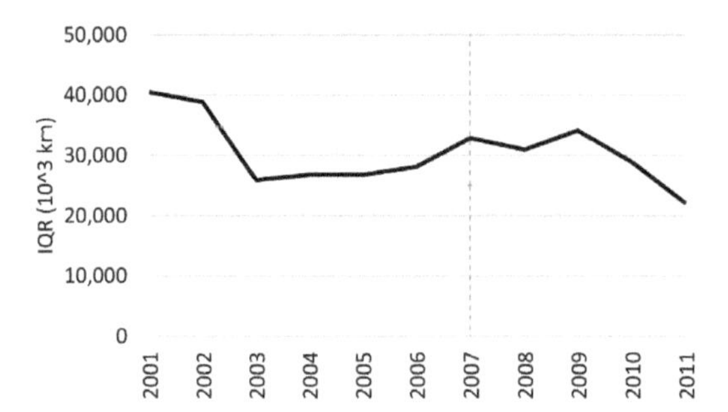

Figure 6. Interquartile range (IQR) of stock evolution. Note: dashed line represents the starting year for the implementation of the stock optimisation strategy.

Normalisation of Indicators

To better understand the evolution of stock-flow-service nexus, each interaction (i.e., service-flow, service-stock and flow-stock) was normalised to 2001. An efficient scenario is one where all the efficiency indicators increase through time whilst the maintenance rate and IQR decrease. As Figure 7 shows, and as explained in the previous subsections, the bus company's stock efficiency and fuel efficiency were not optimal because there was no significant improvement. In addition, those indicators linked to waste (i.e., the stock degradation efficiency and the stock maintenance rates) demonstrate poorer performance. To mitigate the issues identified by these indicators, the company would have had to develop partnerships with suppliers or establish waste management strategies that either extended product lifespan (e.g., by using a lubricant that needed to be substituted every 15,000 rather than

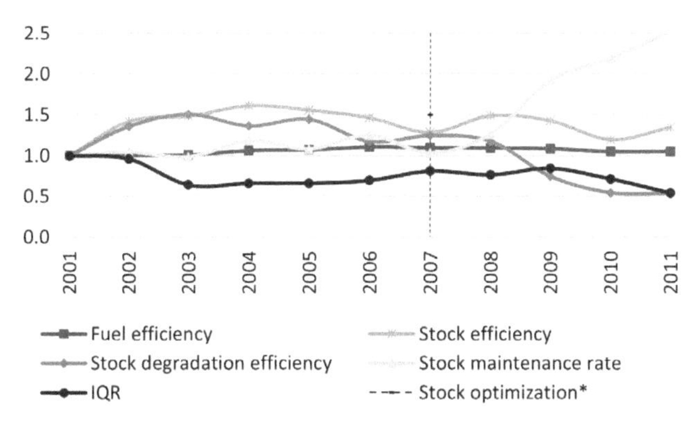

Figure 7. Nexus indicators normalisation. (*) dashed line represents the starting year for the implementation of the stock optimisation strategy.

10,000 km) or which would have led to higher recoverability rates. The lowest level of fluctuation was in fuel consumption. However, in financial terms, given that fuel costs represented 35% of the company's budget, a 1% decrease in efficiency would have led to a 0.4% increase in expenditure (which is equivalent to the annual cost of the urea required by the Selective Catalytic Reduction for NOx reduction purposes in the Euro IV fleet).

Discussion and Concluding Remarks

Through the efficiency indicators explored in this chapter, we have demonstrated that the stock-flow-service nexus approach can be used to quantify the extent at which flows relative to stocks, flows relative to service and stocks relative to service mobilise a bus fleet. In 2011, the bus company required 0.51 kg of diesel fuel and 1.3×10^2 units of vehicle stock to provide 1 km of service. In turn, this 1 km of service led to the degradation/loss and replacement of 2.9×10^2 kg of material flows to ensure stock functionality at the expected standard. This kind of information gets overlooked if a company does not consider the role of stock accumulation, overemphasises fuel efficiency and ignores services units.

The nexus approach allows for a better understanding of where and to what extent a company's physical assets are contributing to strategic goals, such as the optimisation of the in-use fleet (as identified by the IQR) and waste reduction (which is captured by the stock maintenance rate). The nexus can also help corporate decision makers weigh up whether it is best to invest in fuel efficiency, stock efficiency, waste reduction or find a balance between the three. Quantifying interactions between stock (e.g., the vehicle chassis) and material flows (e.g., tyre tread) may also support future procurement strategies where, for instance, a slightly more expensive but well-designed vehicle can be argued for on the basis of fuel and lubricant savings. This is particularly useful when service and operation are affected because certain components must be replaced at a given number of kilometres or certain levels of wear and tear. It is especially beneficial when a company is having to replace parts prior to a bus manufacturer's specified max number of kilometres. The service nexus approach can also support a more solid understanding of operational dynamics, which may permit a more accurate prediction of the fleet or spare parts replacement, thus decreasing the amount of time a vehicle stays in the garage. This perspective also has the potential to provide further evidence for addressing driver behaviour, if that is the root cause for shorter than expected vehicle or part lifecycles or if it represents the only financially viable way to increase profitability and/or improve user experience.

Via the stock efficiency indicator, a management team can identify where material accumulation could provide more services. This may lead to a better allocation of resources where, for instance, parked buses are used for driver training so that new buses are not bought explicitly for that purpose or office spaces are co-shared. In terms of waste management, decision makers could use the nexus to justify vehicle lifetime extension via the conversion of buses into temporary offices, mobile libraries, museums or driver classrooms. The nexus approach could also help companies to establish evidence-based targets. For instance, it can identify which waste outflows are coupled with kilometres and which ones are associated with corrective maintenance due to road accidents or low-quality spare parts. With this knowledge, one can propose a reasonable waste reduction target that aligns with data rather than policies that respond solely to public sentiments and marketing goals.

In short, once a company reaches a product-focused performance plateau, the stock-flow-service nexus can offer a nuanced perspective that changes the way corporate leaders and executive boards consider resources and wastes. This is particularly the case if they are not aware of the importance of material stocks, how they drive flows (and vice versa) and by extension, help to determine service quality. The service nexus could also be used to ascertain the sustainability of a transition from conventional to electric/hybrid vehicles. This is because it can identify the extent to which stock efficiency and fuel efficiency are affected by changes in the material composition of the vehicle. According to Grütter (2014), a 12-metre electric bus that is able to travel 200 km between charging cycles contains a battery weighing around 3 tonnes. This mass increase will, due to axle-weight restrictions, reduce the number of passengers allowed to embark on any one trip. A move towards electrified bus fleets is, therefore, something that needs to be carefully considered on transport systems as busy as Bogota's BRT.

There are various challenges that would need to be overcome before the nexus approach, the concept of material services and its associated indicators are widely accepted. There may be reluctance among key stakeholders to use yet another measuring stick because of the cost and effort involved. In addition, there often needs to be overwhelming evidence that the metric matters to a company and its shareholders or a government and members of the public. Another challenge is the integration of the various aspects (safety, continuity, cleanliness, user satisfaction, accessibility, etc.) that contribute to mobility as a service, given that they are measured using different units, which do not necessarily lend themselves to material allocation. For instance, how should one link vehicle stock to cleanliness? Neither is it clear how or to what extent steel supports hygiene. In addition, when evaluating multicriteria performance, it can be difficult to prioritise actions, even if a weighting method is applied. Weighting can be a subjective exercise and may reflect corporate interests and not those of the customers a company is supposedly serving. Such issues need to be fully considered if one is to make the most of the advantages that a nexus perspective offers.

Acknowledgements

L.G.C. acknowledges the financial support of Fundação para a Ciência e a Tecnologia (FCT) and MIT Portugal Program through the grant PD/BD/128038/2016. L.G.C. acknowledges the support of Colciencias. K.W. acknowledges the financial support of UCLouvain through the FSR Post-doc 2020 fellowship and by the Fund for Scientific Research (FNRS).

References

Ally, J. and Pryor, T. 2007. Life-cycle assessment of diesel, natural gas and hydrogen fuel cell bus transportation systems. Journal of Power Sources, 170, 401–411.

Barrett, P., Sexton, M. and Curado, M. 1998. Sustainability through integration. pp. 1767–76. In Proceedings, CIB World Building Congress 1998, Symposium D, Gävle, Sweden.

Bocken, N. M., Short, S. W., Rana, P. and Evans, S. 2014. A literature and practice review to develop sustainable business model archetypes. Journal of Cleaner Production, 65, 42–56.

Boiral, O. 2011. Managing with ISO systems: lessons from practice. Long Range Planning, 44, 197–220.

Boiral, O., Guillaumie, L., Heras-Saizarbitoria, I. and Tayo Tene, C. V. 2018. Adoption and outcomes of ISO 14001: a systematic review. International Journal of Management Reviews, 20, 411–432.

Brand-Correa, L. I. and Steinberger, J. K. 2017. A framework for decoupling human need satisfaction from energy use. Ecological Economics, 141, 43–52.

BRTdata.org. 2019. Annual demand (passengers per year).

Brundtland, G. H. 1987. Report of the World Commission on environment and development: "Our common future." New York, NY, USA: United Nations.

Carmona, L. G. and Ocampo, A. I. 2014. An evaluation of the environmental performance of bus rapid transit and the contribution to public policy in emerging cities using bogota as a case study. pp. 107–145. *In*: Cotte Poveda, A. and Pardo Martinez, C. I. (eds.). Health, Violence, Environment & Human Development in Developing Countries. New York, US: Nova Science Publishers Inc.

Carmona, L. G., Whiting, K., Carrasco, A., Sousa, T. and Domingos, T. 2017. Material services with both eyes wide open. Sustainability, 9, 1508.

Carmona, L. G., Whiting, K., Haberl, H. and Sousa, T. 2020. The use of steel in the United Kingdom's transport sector: a material stock-flow-service nexus case study. Journal of Industrial Ecology. https://doi.org/10.1111/jiec.13055.

Cervero, R. and Dai, D. 2014. BRT TOD: Leveraging transit oriented development with bus rapid transit investments. Transport Policy, 36, 127–138.

DANE. 2018. Informe comité nacional de expertos para la evaluación del censo nacional de población y vivienda de Colombia 2018 (in Spanish). Bogotá, Colombia: Departamento Administrativo Nacional de Estadística.

Day, R., Walker, G. and Simcock, N. 2016. Conceptualising energy use and energy poverty using a capabilities framework. Energy Policy, 93, 255–264.

Fell, M. J. 2017. Energy services: A conceptual review. Energy Research & Social Science, 27, 129–140.

Fishman, T., Schandl, H., Tanikawa, H., Walker, P. and Krausmann, F. 2014. Accounting for the material stock of nations. Journal of Industrial Ecology, 18, 407–420.

Grütter, J. 2014. Real world performance of hybrid and electric buses. Grütter Consulting.

Guinee, J. B., Heijungs, R., Huppes, G., Zamagni, A., Masoni, P., Buonamici, R., Ekvall, T. and Rydberg, T. 2010. Life cycle assessment: past, present, and future. ACS Publications.

Gunningham, N. and Sinclair, D. 1999. Environment management systems, regulation and the pulp and paper industry: ISO 14001 in practice. Environmental and Planning Law Journal, 16, 5–24.

Haberl, H., Wiedenhofer, D., Erb, K. H., Görg, C. and Krausmann, F. 2017. The material stock–flow–service nexus: a new approach for tackling the decoupling conundrum. Sustainability, 9, 1047.

Hensher, D. A. and Golob, T. F. 2008. Bus rapid transit systems: a comparative assessment. Transportation, 35, 501–518.

Hidalgo, D. 2019. Celebrating 18 Years of TransMilenio: Growing Pains and What Lies Ahead for Bogotá's BRT. The City Fix.

ISO. 2015a. ISO 9001:2015 Quality management systems: Requirements. Geneva, Switzerland: International Organization for Standardization.

ISO. 2015b. ISO 14001:2015 Environmental Management Systems: Requirements with Guidance for Use. Geneva, Switzerland: International Organization for Standardization.

Kalt, G., Wiedenhofer, D., Görg, C. and Haberl, H. 2019. Conceptualizing energy services: A review of energy and well-being along the energy service cascade. Energy Research & Social Science, 53, 47–58.

Krausmann, F., Weisz, H. and Eisenmenger, N. 2016. Transitions in sociometabolic regimes throughout human history. pp. 63–92. *In*: Haberl, H., Fischer-Kowalski, M., Krausmann, F. and Winiwarter, V. (eds.). Social Ecology: Society-Nature Relations across Time and Space. Cham, Switzerland: Springer.

Krausmann, F., Wiedenhofer, D., Lauk, C., Haas, W., Tanikawa, H., Fishman, T., Miatto, A., Schandl, H. and Helmut Haberl. 2017. Global socioeconomic material stocks rise 23-fold over the 20th century and require half of annual resource use. Proceedings of the National Academy of Sciences, 114, 1880–1885.

Krausmann, F., Lauk, C., Haas, W. and Wiedenhofer, D. 2018. From resource extraction to outflows of wastes and emissions: The socioeconomic metabolism of the global economy, 1900–2015. Global Environmental Change, 52, 131–140.

Liyin, S., Hong, Y. and Griffith, A. 2006. Improving environmental performance by means of empowerment of contractors. Management of Environmental Quality: An International Journal, 17, 242–257.

Müller, D. B. 2006. Stock dynamics for forecasting material flows—Case study for housing in The Netherlands. Ecological Economics, 59, 142–156.

Pauliuk, S. and Müller, D. B. 2014. The role of in-use stocks in the social metabolism and in climate change mitigation. Global Environmental Change, 24, 132–142.

Reis, A. V., de O. Neves, F., Hikichi, S. E., Salgado, E. G. and Beijo, L. A. 2018. Is ISO 14001 certification really good to the company? a critical analysis. Production, 28.

SDP. 2018. Análisis demográfico y proyecciones poblacionales de Bogotá (in Spanish). Bogotá, Colombia: Alcaldía Mayor de Bogotá, Secretaría Distrital de Planeación.

Simonsen, M. 2012. Energi- og utslippsvirkninger av produksjon av Volvo 8500 busser (in Norwegian). Sogndal, Noruega: Vestlandsforsking.

Steger, U., Wei, L. and Zhaoben, F. 2017. Greening Chinese Business: Barriers, Trends and Opportunities for Environmental Management. 2nd ed. Abingdon, UK: Routledge.

Transmilenio. 2017. Estadísticas de oferta y demanda del Sistema Integrado de Transporte Público - SITP - Octubre 2016 (in Spanish).

Transmilenio. 2019. Historia de TransMilenio (in Spanish).

UNFCCC. 2015. AM0031: Large-scale Methodology, Bus rapid transit projects, Version 06.0, Sectoral scope(s): 07. Rio de Janeiro, Brasil and New York, US: United Nations Framework Convention on Climate Change.

Weidema, B. P. 2010. Environmental training for the food industry. pp. 356–373. In Environmental Assessment and Management in the Food Industry. Oxford, UK: Elsevier.

Weidema, B. P. 2018. The social footprint—a practical approach to comprehensive and consistent social LCA. The International Journal of Life Cycle Assessment, 23, 700–709.

Weisz, H., Suh, S. and Graedel, T. E. 2015. Industrial ecology: The role of manufactured capital in sustainability. Proceedings of the National Academy of Sciences, 112, 6260–6264.

Whitelaw, K. 2012. ISO 14001 Environmental Systems Handbook. Oxford, UK: Elsevier.

Whiting, K., Carmona, L. G., Brand-Correa, L. I. and Simpson, E. 2020. Illumination as a material service: A comparison between Ancient Rome and early 19th century London. Ecological Economics, 169C, 106502.

Wiedenhofer, D., Fishman, T., Lauk, C., Haas, W. and Krausmann, F. 2019. Integrating material stock dynamics into economy-wide material flow accounting: concepts, modelling, and global application for 1900–2050. Ecological Economics, 156, 121–133.

World Bank. 2019. Gross fixed capital formation in % of GDP.

Wright, L. and Hook, W. 2007. Bus rapid transit planning guide. New York, US: Institute for Transportation & Development Policy.

Wulf, C., Zapp, P., Schreiber, A., Marx, J. and Schlör, H. 2017. Lessons learned from a life cycle sustainability assessment of rare earth permanent magnets. Journal of Industrial Ecology, 21, 1578–1590.

CHAPTER 4

Land Tenure, Ethnicity, and Desertification in Darfur#

Osman Suliman

Introduction

With the onset of the recent political turmoil in Sudan, the debate on possible solutions to peripheral conflicts has been revived. Important among these is the Darfur conflict. Darfur is a land-locked, remote area in Western Sudan. It is about the size of France, approximately 250,000 square kilometers in area (Figure 1). Droughts and desertification have contributed significantly to the underdevelopment of Darfur, especially northern Darfur. Darfur experienced severe droughts in 1973–74 and again in 1984.[1] The resulting continuity of conflict over land tenure (property rights) and food insecurity has displaced internally and externally over 1.75 million.[2] At the beginning of the conflict, these attacks involved both the government of Sudan (GOS) and the Janjaweed, a proxy militia, presumably with the purpose of flushing out the rebels.[3] With the progression of the conflict, militias have seized the opportunity of increased lawlessness to loot, rape, and kill with impunity.[4]

The population of Darfur grew from about one million in 1956 to over six million in 2015, an increase of 84%. While the population has been exponentially increasing and with the expanding insecurity in Darfur, production of the main food crop in Darfur, sorghum, tumbled significantly after the onset of the conflict (Tables 1 and 1a). Coupled with the declining rainfall and subsequent droughts, this substantial increase in population exacerbated the exhaustion of natural resources and the concomitant drop in productivity and livelihoods (Table 1 and Figure 2). The rise in

Millersville University, Economics Department, P.O. Box 1002, Millersville, PA 17551, USA.
Email: osman.suliman@millersville.edu

An earlier version of this chapter was prepared for The Sustainability and Development Conference, University of Michigan, Ann Arbor, October 2019.

[1] de Waal (1989).

[2] Bremer et al. (2006).

[3] Ibid.

[4] Ibid.

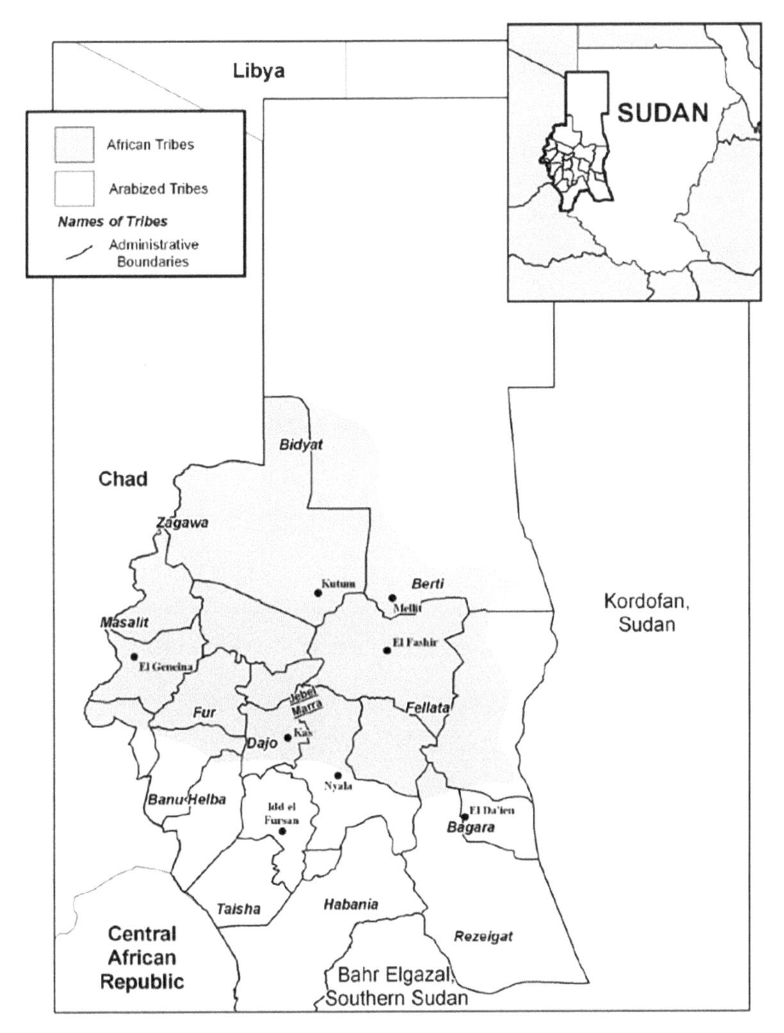

Figure 1. Map of Darfur region.

population density necessitated an expansion of the cultivated area and the resulting disruption of the fallow/cropping rotation. As Fadul (2004) points out, to minimize the risk of decreasing crop productivity, farmers tried to diversify by shifting to small animal herding, particularly goats and sheep. This increased the animal population and decreased the carrying capacity of the natural grazing lands.

In the arid regions of Darfur, farmers have long relied on a system of rotation and intercropping, producing both cereal crops and gum arabic from Acacia trees. However, this system is breaking down due to pressure from droughts, desertification, population increase, and large-scale mechanized agriculture (Table 2). According to the UNEP (2007), the most environmentally damaging aspects of government policy have been the promotion of rain-fed mechanized agriculture, the subsequent institutional failure to assess its negative consequences when they first became

Table 1. Areas and productivity of sorghum in Darfur.

	All of Darfur		**Northern Darfur***		
	Sorghum	**Sorghum**	**Sorghum**	**Sorghum**	
	Cultivated area	**Productivity**	**Cultivated area**	**Productivity**	
Year	**Feddans**	**Kg/Feddan**	**Feddans**	**Kg/Feddan**	**Population**
1970–71	213	512	13	200	
1971–72	186	338	16	100	
1972–73	153	381	20	200	
1973–74	486	152	38	100	1,340,000
1974–75	383	255	59	220	
1975–76	422	254	127	337	
1976–77	448	274	65	116	
1977–78	572	436	90	135	
1978–79	574	286	85	192	
1979–80	425	196	69	49	
1980–81	536	220	60	161	
1981–82	534	270	60	188	
1982–83	640	214	180	154	
1983–84	290	96	52	87	3,500,000
1984–85	454	119	40	56	
1985–86	478	213	75	122	
1986–87	730	88	262	67	
1987–88	559	52	36	40	
1988–89	748	199	68	250	
1989–90	371	68	62	69	
1990–91	812	83	57	77	
1991–92	740	256	48	250	
1992–93	1,060	225	70	218	
1993–94	613	181	74	174	5,600,000
1994–95	976	338	82	214	
1995–96	724	264	10	15	
1996–97	337	309	67	87	
1997–98	1,278	359	93	115	
1998–99	1,003	303	76	230	
1999–2000	1,300	223	53	98	
2000–01	1,409	222	156	150	
2001–02	1,933	285	90	143	
2002–03	1,852	209	105	130	
2003–04	1,865	236	89	120	6,480,000
2004–05	955	184	53	95	

Table 1 Contd. ...

...Table 1 Contd.

Year	Feddans	Kg/Feddan	Feddans	Kg/Feddan	Population
2005–06			147	178	
2006–07			89	60	

Source: Ministry of Agriculture and Forests, Department of Agricultural Statistics, Time Series of Area Planted, Harvested, and Yield Data of the Main Food and Oil Crops by States and Mode of Irrigation 07/1971–04/2005, Volume 1, September 2006.
* Desertification is higher in northern Darfur.

Table 1a. Northern Darfur sorghum productivity.

Mean	140.4594595
Standard Error	11.65614678
Median	130
Mode	200
Standard Deviation	70.90157285
Sample Variance	5027.033033
Kurtosis	0.115459264
Skewness	0.537610538
Range	322
Minimum	15
Maximum	337
Sum	5,197
Count	37
Coefficient of Variation = 0.504783182	

Source: Based on data from Table 1.

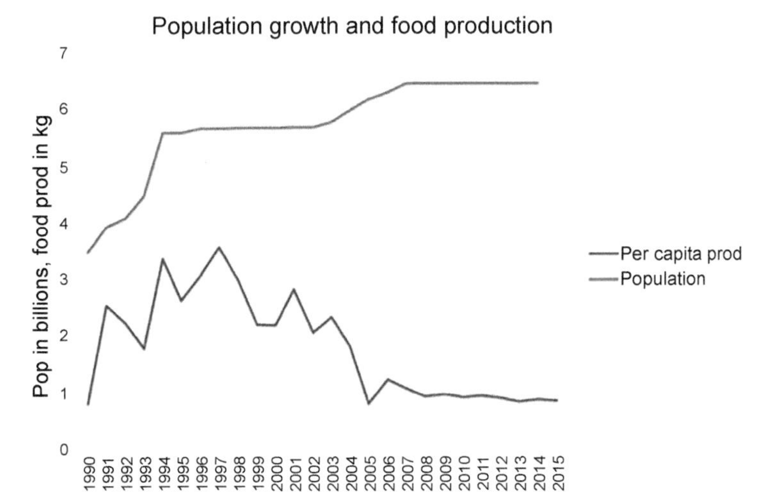

Figure 2. Population growth and food production.

Table 2. Changes in rangeland coverage.

Study area and state	**Original and final pasture land**	**Annual linear rate + (period loss)**	**Comments**
Jebel Marra	2.9 to 23.0 from 1973 to 2000	+ (289%)	Increase in open forest land, decrease in closed forest and bush and shrubland
Timbisquo South Darfur	65.4 to 59.3 from 1973 to 2000	– (9.3%)	Loss due to the expansion of mechanized agriculture, bush and shrubland, flood and wetland
Um Chellata South Darfur	42.4 to 32.7 from 1973 to 2000	– (65%)	Loss due to the expansion of mechanized agriculture, increase in degraded areas and flood land, decrease in a grassland area
Darfur		N/A	No simple trend. Jebel Marra anomalous, South Darfur similar to South Sudan with agricultural expansion

Source: UNEP (2007)

clearly apparent, and the lack of governance in the area of pesticide management, leaving the country with a difficult and expensive environmental legacy.

The Darfur crisis emanated from poor governance that triggered a series of endogenous interactions of socioeconomic institutional failures and geography (Figure 3). Many of the studies of the Darfur conflict are based on the premise that it is mainly a land tenure/use or property rights problem. In the survey that I conducted in 2007–08, 49% of the households and focus groups argued that land tenure is the main proximate cause of the conflict, while 29.3% thought it is desertification, the ecological shrinkage of grazing vegetation, and the resulting trespassing of nomadic Arab pastoralists to sedentary farms owned mainly by African groups. The conflict also has deep roots within Darfur and 17.7% of the respondents thought the conflict grew out of ethnic tensions regarding native administration appointments.[5] Native administration is responsible for adjudicating local law and order of society.

In the literature, there is strong support for the land-tenure hypothesis. Unruh and Abdul-Jalil (2014), Abdul-Jalil and Unruh (2013), and Abdul-Jalil (2006) argued that land tenure in Darfur acted as a principal protagonist to the cause of the ongoing armed conflict. Olsson and Siba (2013) and Olsson (2010) suggested that capturing access to water and land quality appears to have played an important role in the conflict. For geography, Kevane and Gray (2008) asserted that data on rainfall patterns weakly corroborate the hypothesis that climate change explains the Darfur conflict. Hibbs and Olsson (2004) contend that in historical times, geography and biogeography are prime forces of current prosperity, even though the quality of institutions has a strong connection to national levels of prosperity. The hypothesis of ethnic tensions is supported by Moscona et al. (2017), Vanrooyen et al. (2008), and

[5] See Suliman (2011), Chapter 3.

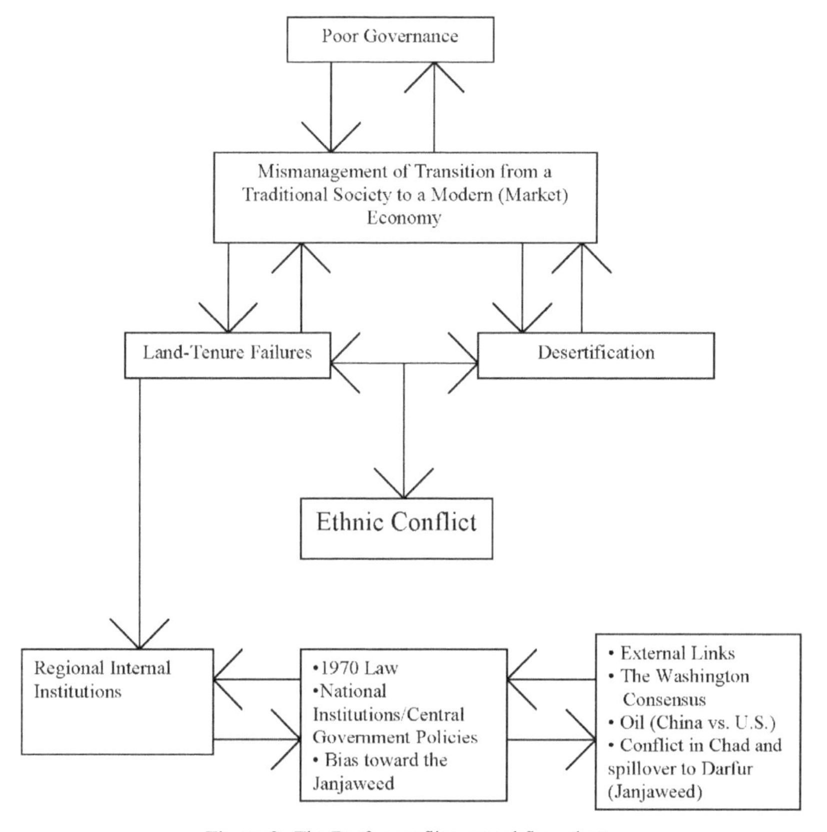

Figure 3. The Darfur conflict: causal flow chart.

Ahlerup and Olsson (2012). Moscona et al. proved the hypothesis that ethnic groups that are organized around segmentary lineages are prone to violence and conflict.

This chapter attempts to test the above hypotheses by specifying three proxy variables for land tenure, desertification, and ethnic tensions over native administration appointments (law and order).[6] For land tenure, the per capita production of the main food crop, sorghum, is used (X2). As discussed, desertification intensified the competition over expanding acreage of the food crop, together with the expanding population size, and the resulting shrinkage of grazing land for pastoralists. Desertification is represented by the ratio of the number of vegetation species available for grazing to the total acreage of the food crop (X10). The third variable is ethnic tensions over native administration leadership positions to keep local law and order (X7). Stepwise regressions that included other variables were also conducted. As expected, significant multicollinearity existed and the final qualitative regression was run on the three variables.

[6] Native administration is a system of indirect ruling responsible for maintaining law and order and tax collection. Also, to mitigate the possible endogeneity problem of the land tenure variable, per capita production of sorghum, the main food crop, is used as an instrumental variable.

Background

Institutional Failures and Geography

Socially, as Rabbah (1998) posits, notwithstanding their historic standing as one of the largest kingdoms that ruled in Sudan, the survival and continuity of the Fur kingdom emanate from their amiable tendency to promote diversity and intermarry with others (Table 3, Figure 1). Besides their intermarriage with the Arabs of North Africa, the Fur are a mixture of negroid and semi-negroid groups such as the Keira, the Firtit, and the Taqmuurrkwa. As a matter of fact, the ruling family of Sulayman Solongdungo has been said to have some Arabic roots.[7] However, despite the diverse

Table 3. Examples of recent qabilah reconciliation conferences.

Purpose of the Conference	Date	Place	Administrative Unit	Causes of the Conflict
To resolve the Maaliya and Razeigat *qabilah*s	October 2004	Nyala City	Abu-Jabra and Adila	Land and administration
Peaceful and social reconciliation between the Birgid and REzeigat *qabilah*s	January 2005	Nyala City	Asalaya and Yas	Rebellion
Reconciliatory conference between *qabilah*s at the El-Malalam Wahda	February 5, 2005	El-Malalam Wohda	El-Malalam	Rebellion and land
Peaceful co-living conference between El-Misiria and Birgid *qabilah*s	February 16, 2005	Nyala City	Nitaiga and Shiaria	Rebellion
Peaceful resolution between the Habania and REzeigat *qabilah*s	March 2005	Nyala City	Buram, Giraida	Grazing areas
Conference between *qabilah*s living on the border between Sudan and the Republic of Central Africa	March 2006	UmDafuq Area	UmDafuq, Republic of Central Africa	Grazing areas
Reconciliation conference between the Ziltain, Fur, Birgid, El-Sida, Rezeigat, and Burnu *qabilah*s	February 2005	Nyala City	Nyala	Rebellion
Reconciliation conference between *qabilah*s living in the east and south of Nyala: Dajo, Fur, Zaghura, Turjum, and Hotia	February 2005	UmDafuq	Abu-Ajura	Rebellion
Kas conference for reconciliation	April 2006	Kas locality	Kas	Land and administration

Source: Takana (2007), El-Ayam Daily

[7] Rabbah (1998), p. 20.

nature of the Fur, their kings worked ardently to amalgamate this ethnicity into a bifurcated society with one branch belonging to the Keira and the other to Kunjaara.[8] Because of this social integration, the Fur, mainly cultivators, co-lived peacefully with the pastoralist Arabs, with mutually beneficial backward/forward patterns of production, where pasturing improved fertility of the soil and animals ate crop remnants after harvest. As such, it was in the interest of both parties to resolve their differences wisely through local mediators (Table 3).

According to Rabbah (1998), even with the willingness to integrate, the psychology of the Arabs and the Fur remained different. Relatively, the Fur are more educated and cultured because of their sedentary style of living. However, as Rabbah (1998) posits, being discriminated against as negroid Nuba might have created an inferiority complex among the Fur. As noted by Ibn Khaldoun (no date), the Arabs, on the other hand, are less educated due to their nomadic life with a notable proclivity for violence and are perceived by the Fur as savages.

Desertification conditions have caused pastoralists in northern Darfur to migrate south, seeking better pastures and water (Table 1). In turn, low-level conflicts escalated into state-wide conflicts, as these pastoralists began to trespass on sedentary farmers' properties, one of the major institutional failures and poor governance consequences. The country failed to develop the necessary infrastructure for improved climate analysis, disaster predictions, and risk reduction. The key issue in Darfur and the country as a whole is that environmental degradation and desertification are shrinking the resource base, while the population is increasing demands on these resources. In Darfur, the UNEP (2007) and the follow-up Darfur Joint Assessment Mission observed an active lumber industry and lack of sustainable natural resource management in conflict areas. Al Mangouri (2004) points out that institutions which have been used to restrict desertification in other parts of the country under the Master Plan and Desert Encroachment Control and Rehabilitation Program (DECARP) through the joint work of the Ministry of Agriculture, the Food and Agriculture Organization, the UNEP, and the U.S. National Research Council did not operate in this area, despite its significance to gum arabic production. An environment assessment by the UN Office for the Coordination of Humanitarian Affairs (OCHA), conducted at three camps in Darfur in 2001, emphasized that there is no international agency with a specific mandate to incorporate environmental issues into peace efforts.

In Darfur, severe deforestation occurred because of the problem of fuelwood and timber provision. This also contributed to a major human rights abuse as women from the camps are often attacked and raped when they leave camps to search for fuelwood. The UNEP (2007) reports that Darfur lost more than 30% of its forest since 1956, mainly because of deforestation resulting from fuelwood and charcoal extraction, mechanized agriculture, traditional rain-fed, shifting agriculture, drought and climate change, overgrazing and fires, direct conflict impacts, commercial lumber, and to a lesser extent the export industry and traditional construction (Table 4).

Another problem is poor water resources management. Traditional rainwater runoff dugouts have been critical in supplying water for both domestic and pastoralist

[8] Ibid.

Table 4. Summary of deforestation rates in Darfur, 1973–2006.

Study area and state	Original and final forest and woodland cover	Annual linear rate + (period loss)	Comments
Jebel Marra	50.7 to 35.8 from 1973 to 2000	1.04% (29.4%)	Closed forest changing to open forest land and burnt areas
Timbisquo South Darfur	72.0 to 51.0 from 1973 to 2000	1.33% (29.1%)	Closed forest and wooded grassland replaced by burnt areas and rain-fed agriculture
Um Chellata South Darfur	23.8 to 16.1 from 1973 to 2000	1.20% (32.4%)	Closed forest replaced by burnt areas, pasture, and rain-fed agriculture
Darfur		1.19% (30.3%)	Rapid and consistent deforestation approximately one-third complete by 2006

Source: UNEP (2007)

use in remote areas vulnerable to erratic rainfall variations. Over time, topsoil erosion, drifting sands, and poor maintenance have led to a serious decline in water storage capacity and loss of many dugouts.

Large-scale mechanized agriculture to the north and west has pushed traditional agriculture to southern Darfur. It has also taken most of the land previously used by pastoralists.

Clearly, the existing institutional structure is too weak to resolve the onerous problem of gun control and security in rural Darfur, compensatory development projects, funding for refugee losses, rudimentary financial institutions that mobilize investors, and civil democratic institutions. Much of the conflict arose as a result of lack of security, exponential population growth, the paucity of investment planning to keep up with this growth, and social factors. As such, there are internal (regional and local) factors together with external factors that led to these institutional failures.

The regional internal factors include land control and rent-seeking issues, environmental degradation problems, the marginalization of the people of Darfur, and the effects of market restrictions and the war on the economy of the region (Figure 3). Young and Osman (2006) point out that in some rural areas the government of Sudan (GOS) controls the corridor extending from western Darfur up into north Darfur toward Kutum where there are higher concentrations of Arab groups which control access to natural resources. Issues may include land occupation and permitting livestock to graze the fields of neighboring farmers, thereby destroying their crops. In addition, the Arab groups acquire profits (rent-seeking) by forcing protection payments from local residents, which feeds directly into genocide accusations by international human rights groups.

Another important failure has been in the management of land tenure. Traditional community-based land management systems (*hawakir*) were reasonably effective. This situation was radically changed in the 1970s by a number of ill-planned initiatives. The 1970 land-tenure law failed because it ignored the existing informal institutions and customary laws in Darfur.

Land tenure in Sudan is problematic and a major obstacle to sustainable land use. Prior to the 1970s, communal title to shared rural land was generally acknowledged at the local level but undocumented. The 1970 law was doomed to failure because it completely ignored informal institutions and customary laws of Darfur (Figure 3). The 1970 People's Local Government Act marginalized the pre-existing traditional land management systems, which were providing vital judicial checks and balances (*judiya*) in the absence of a modern land tenure system. Abdul-Jalil (2006) recommends the formation of a land commission to establish and implement a land-use policy for development and registration of land that encompasses historical development, informal institutions, and customary laws in Darfur.

The marginalization of Darfur and the central control of wealth and power were discussed by the Darfur Peace Agreement (DPA) in Abjus, Nigeria, the seven commissions of the Ahl al-Sudan initiative, the Doha initiative, the African Union Panel in Darfur, and several local reconciliation conferences on land and grazing areas (Table 3). As Young and Osman (2006) report, this sense of marginalization was felt by Arabized (and some African) groups which may have in part been a reason why they joined the government counterinsurgency and participate in what is seen as a wider 'Arab movement'. While the humanitarian community has addressed these disparities in 2005, Arabized groups were excluded and lacked proper representation at the DPA, fostering further grievances. While this rift between the two groups emanated from competition over property (land) rights, pastoral areas, and livestock routes, it has been abysmally abused by the NCP (1989–2019) and Sadek el-Mahdi (1985–1989) governments.[9]

The conflict debilitated trade in all Darfurian commodity markets. This resulted in reducing trade between primary and secondary markets with more producers serving only local markets. As such, total assets (other than livestock and land) fell by about 69% from an average of $30,895 to an average of $9,500.[10] In the past, livestock marketing was one of the mainstays of the economy, but with the advent of the conflict, its demise has affected livelihoods across Darfur, including an estimated 575,000 pastoralists and a high proportion of agro-pastoralists. On average, the estimated loss in livestock assets is 54%. Also, a high proportion of traders have been forced out of business and left Darfur, including the traders in cash crops such as tobacco and groundnuts. Some markets have also suffered a loss of physical infrastructure and the closure of customs offices. The decline in trading activity has an effect on the availability of credit for farmers and falling prices.[11]

Institutional failures were exacerbated by the privileged access of the National Conference Party (NCP) members to public resources. Clientelism coupled with low state capacity and the ideological preferences of decision-makers interacted to influence the policy status quo. As de Walle (2001) points out, these interacting variables are largely responsible for the initial resistance of policymakers to any kind of reform at the outset of the regime change. Later, as the NCP did, a mixture of partial reform and politically manipulated reform were pursued. The ideological

---

[9] Young and Osman (2006).

[10] Ibid.

[11] Ibid.

predilections of policy elites were often overlooked in explaining the course of economic reform. The NCP's version of Islamic Sharia law applied to a partisan rather than national agenda, the local attachment to planning, and their concern for equity issues as anachronisms are all retained as disingenuous use of ideology for political purposes. There was a tendency to delay reform policies until they are politically feasible leading to the persistence of inflationary policies, such as floating of the Sudanese currency, and permitting a multitude of unfair foreign investments in the real estate and agricultural schemes sectors. They continued deficit-financing policies and international borrowing allowing the international debt to rise to $28.4 billion at the end of 2006, which was 77% of GDP and 86% of total exports.

Livelihood Impact of the Conflict

Based on the survey (of 500 households) and focus group data gathered for this research, about 70% of the surveyed population of Darfur was living in camps, 8% were living with other family members, 13.5% were renting houses, and 8% were roaming as homeless.[12] Of those surveyed, in 28% of the families, women were the head of the household. In terms of domicile, 68.8% of the families were rural, 9.1% were nomads, and 21.1% came from urban areas. Of those, 35.2% identified themselves as farmers, 17.9% as pastoralists, 6.9% as merchants, 12.8% as professionals, and 27.2% as having other careers. However, only 20.6% of these groups have steady income sources, and only 4% receive income support from other family members (Table 5).

Asset losses can be assessed by comparing the value of ownership before and after the conflict (Tables 6–10). While 85.3% of the surveyed Darfurians claimed ownership of some assets (animals and land) before the conflict, only 11% of those did not lose any of their assets. Table 6 shows that before the conflict, 68.8% of the surveyed households owned homes with an average value of $9,000 U.S. dollars. About 32% of the respondents indicated that they lost their homes. After the conflict only 19.2% of the total respondents reported that they owned homes, and the average value of their homes increased to $10,000 U.S. dollars.

Table 5. Current annual income (in millions of U.S. dollars).

	N	Minimum	Maximum	Mean	Std. Deviation
Annual individual fixed income	103 (20.6%)	0.00	1.00	0.01	116.97
Annual income support from other members of the family	20 (4%)	0.00	0.50	0.03	133.56

Notes: 1. 1 U.S. dollar (USD) = 2 Sudanese pounds. 2. Total sample size = 500 thousand. 3. N = number of responding households. Numbers in parenthesis are the percentage of responding households out of the total sample. 4. The high standard deviation and range between the minimum and maximum levels of income indicate that there is high inequality in income distribution. That also shows that the median income is significantly lower than what the mean reflects.
Source: Suliman (2011)

[12] The results for this survey are published in Suliman (2011), Chapter 3.

Table 6. Residential housing (in millions of U.S. dollars).

	Sum of Total Value of Houses for All Respondents*	**Mean Value**
Before the conflict N = 344 (68.8%)	3.2	0.009
After the conflict N = 96* (19.2%)	.93	0.01

Notes: 1. Total sample size = 500 households.
Source: Suliman (2011)
* Out of the total respondents, only 96 reported value for their homes. Others may have lost their homes.

Table 7. Agricultural land.

	Area (in Mukhamas)		**Sum of total Value for all respondents (in millions of USD)**	**Mean value (in millions of USD)**
	Total	**Mean**		
Before the conflict N = 327 (65.4%)	41,8795	1,797.40	21.04	0.06
After the conflict N = 51 (10.2%)	366.50	18.33	4.10	0.08

Notes: 1. See notes under Table 3.2. 2. 1 mukhama = 0.60 hectares. 3. The mean value of agricultural land owned per person increased after the conflict from about sixty thousand USD to eighty thousand USD, but the number of people claiming land value decreased.
Source: Suliman (2011)

Table 8. Livestock.

	Headcount		**Sum of the total value for all respondents (in millions of USD)**	**Mean value (in millions of USD)**
	Total	**Mean**		
Before the conflict N = 364 (72.8%)	14,849.00	57.11	11.75	0.032
After the conflict N = 65 (13%)	2,586.00	56.21	8.40	0.129

Notes: 1. See notes under Tables 3.2 and 3.3. 2. The mean value of livestock owned per person rose after the conflict from about 32 thousand USD to 129 thousand USD because the headcount for livestock decreased.
Source: Suliman (2011)

Table 9. Rental income from land.

	Total sum for all respondents (in millions of USD)	**Mean (in millions of USD)**
Before the conflict N = 90 (18%)	.165	0.002
After the conflict N = 34 (6.8%)	0.001	0.000

Source: Suliman (2011)

Table 10. Area of land sold.

	Sum of the total area for all respondents (in mukhamas)	**Mean area (in mukhamas)**
Before the conflict N = 65 (13%)	397.00	6.10
After the conflict N = 6 (1.1%)	14.00	2.33

Source: Suliman (2011)

Data

The data on production of sorghum, the main food crop in Darfur, and population size has been obtained from the Ministry of Agriculture, Department of Statistics (2016). The data on vegetation has been published by Fadul (2004) and statistical publications of the Department of Environment and Desertification in EL Fashir, Darfur (2016).[13] The rest of the socioeconomic data are from the International Country Risk Guide (2018).

Empirical Analysis

Definition of Variables

Dependent Variable: Conflict = y, a binary variable with zero value for the period before the conflict (1990–2002) and one for the conflict period (2002–2015).

Independent Variables:

X1 = vegetation in grazing areas (number of species of grazing grass)
X2 = production per capita (kg)
X3 = population size (millions)
X4 = interaction variable, X1*X2 (institutions * geography)
X5 = government stability
X6 = socioeconomic conditions
X7 = law and order
X8 = ethnic tensions
X9 = interaction variable, X1*X8 (production growth * ethnic tensions)
X10 = ratio of grazing vegetation to areas of cultivated food crop

Descriptive Statistics

Tables 1 and 11 show descriptive statistics. The table shows that the mean per capita production of sorghum, the main food crop in Darfur, is only 0.06 kg per year. This reflects the continuous decline in productivity, while cultivated areas continued to expand because of desertification. Meanwhile, the population size has

[13] This has been corroborated by personal interviews, Mr. Fadul, and the Director of the Department of Environment and Desertification (2008). In 2007–08, I did survey questionnaires and focus groups in Darfur.

Table 11. Descriptive statistics.

Variable	Obs	Mean	Std. Dev.	Min	Max
X1	26	3.915385	3.443219	1	12
X2	26	.0596538	.0252253	.012	.09
X3	26	5.760385	.8494397	3.5	6.48
X4	26	.1762308	.1160706	.058	.473
X5	26	8.157308	3.00233	1	10
X6	26	2.214615	.7886557	1	4.5
X7	26	2.198846	.4547907	1	3
X8	26	.8992308	.7858138	.03	1.97
X9	26	.0648338	.0632587	.00054	.16548
X10	26	.0035	.0039421	.001	.02

been exponentially rising from about one million in 1970 to over six million in 2015 (Figure 1). This combination of lower productivity and an increase in population size led to several famines in Darfur in 1973–74, 1984, and more recently with the onset of the conflict in 2003. As hypothesized in this study, the expansion of cultivated areas and the concomitant erosion of grazing vegetation areas are the main driving forces behind the land-tenure/use problem that led to the conflict. Based on the survey that I conducted in 2007–2008, 49% of the people interviewed thought the main cause of the conflict was the land-tenure problem. Before the advent of desertification, there were plenty of valleys and sedentary farming areas. Farmers followed a shifting cultivation rotation with fallow areas. Pastoralists had plenty of vegetation and grazing land and they did their seasonal north-south migrations through well-defined routes. With desertification and shrinkage of grazing areas, pastoralists started trespassing in sedentary farming areas and conflict erupted. Several mediation conferences between farmers and pastoralists tried to resolve the land-use problem without much success (Table 3).

The interaction variable between cultivated land acreage and vegetation in grazing areas represents an interaction of traditional customary laws of land distribution, as the existing institution of land tenure and geography (X4). But, as the correlation matrix (Table 12) shows, it is negatively correlated with the production per capita variable, indicating that the tradeoff between geography (desertification) and institutions impacted negatively on production per capita (wellbeing).

The government stability with a mean of 8.1 and a standard deviation of 3.0 has a coefficient of variation of 38% and a wide range. The index for socioeconomic stability is also low with a mean of only two with the highest value of 12.

The indexes for corruption, military in politics, religion in politics, and law and order are low mostly in the range of zero to two out of a maximum of six. Lower scores reflect a higher political risk. These variables reflect the unstable political environment which made different factions in Darfur rebel against the military government of Sudan.

Table 12. Correlation matrix.

	logx1	logx2	logx3	logx4	logx5	logx6	logx7	logx8	logx9	lox10
logx1	1.0000									
logx2	–0.6973	1.0000								
logx3	–0.8986	0.6878	1.0000							
logx4	0.7276	–0.0156	–0.5952	1.0000						
logx5	–0.7551	0.6595	0.8786	–0.4222	1.0000					
logx6	–0.0567	–0.1137	0.3540	–0.1878	0.2942	1.000				
logx7	–0.5293	0.4579	0.5938	–0.3001	0.7717	0.4615	1.000			
logx8	–0.8878	0.5653	0.6675	–0.6975	0.6076	–0.2291	0.4203	1.000		
logx9	–0.9186	0.7413	0.7366	–0.5719	0.6800	–0.2184	0.4708	0.9727	1.000	
logx10	0.6102	–0.5023	–0.7502	0.3704	–0.7287	–0.3403	–0.4724	–0.5005	–0.5486	1.000

Probit Analysis

The empirical results assess the situation before (1990–2002) and after the conflict (2003–present) qualitatively. A probit model is employed in which the regress and is zero for no conflict or one for conflict. Let $y = 1$ if there is conflict and $y = 0$ for the absence of conflict. The probit model specifies the conditional probability that $y_i = 1$ such that:

$$P[y=1 \mid X_{1i},\cdots,X_{ni},\beta_0,\cdots\beta_n] = \Phi\left(\beta_0 + \sum_{n=1}^{n}\beta_n X_{ni}\right)$$

where $\Phi(\cdot)$ is the cumulative function of the standard normal distribution. β_0 is the constant term, β_n are the partial slope coefficients and measure the change in the estimated probit for a unit change in the probability of the given regressor with other variables held constant. To mitigate the multicollinearity problem, stepwise elimination of variables is performed. A stepwise probit estimate shows the probability from zero to one of the onset of conflict. Table 12 shows the correlation coefficients between the different variables. Multicollinearity is suspected to be a problem if the deletion of one variable noticeably reduced the significance of another variable or if adding a variable made another variable lose its significance. The backward and forward search for the best fit set among the ten variables led to the selection of three variables. As such, the selection process eliminated multicollinearity existing between variables, such as the interaction variable between agricultural land acreage (land use) and vegetation in grazing areas, one of the variables representing interaction between institutions and geography. There is also high correlation between per capita production of the main food crop and population size. High correlations also exist between the socioeconomic variables, government stability, corruption, military in politics, religion in politics, law and order, and ethnic tensions. The only variables that remain significant in the model are the per capita production of the main food crop, law and order (ethnic tensions), and the ratio of the number of species of grazing vegetation to the total acreage of the main food crop. The latter represents the competitive relationship between pastoralism and sedentary farming. This rigorous research resulted in a three-variable model. The existence of multicollinearity indicates that other variables are catalysts rather than initiating factors. For comparison OLS estimates are also presented. OLS represents a linear probability model. OLS is not logically an attractive model because it assumes that $P_i = E(Y = 1 \mid X)$ increases linearly (remains constant) with X. Table 13 displays that ordinary least squares pose several problems in estimating such models. In reality, P_i is expected to be nonlinearly related to X_i.[14] Restricting analysis to the probit model yields the following results:

$$\text{Conflict} = \underset{(1.64)}{0.41X_2} + \underset{(2.21)}{1.81X_7} - \underset{(-1.74)}{0.32X_{10}} + \mu_t$$

Since the probit model uses the method of maximum likelihood, the estimated error term (μ_t) is asymptotic. Therefore, instead of utilizing the t-statistic to evaluate the significance of coefficients, the standard normal statistic (below the respective coefficients) is employed. Further, in a binary regression model, R^2 is less meaningful and is replaced by a comparable measure, pseudo R^2 (0.3676). For the null hypothesis that all the slope coefficients are simultaneously equal to zero, the likelihood ratio (LR) is rejected. That is, all the regressors have a significant impact on conflict in Darfur, as the LR statistic is 13.25, whose probability value is 0.0041, which is very small. All coefficients in the model have the apriori expected signs and are at least significant at the 10% level of significance.

The probit regression results are reported in Tables 14 and 15 with the marginal effects of the main covariate with conflict as the dependent variable displayed in Table 15. Table 15 shows changes in probability of conflict with the three independent variables. It is clear that an increase in production of the main food crop (X2) increases the probability of conflict by .41 (41%). Noticeably, an increase in ethnic tensions over setting local law and order (X10) will increase the probability by 1.81, almost doubling the probability of conflict.[15] An increase in the vegetation of grazing grass (X7) lowers the probability of conflict by .32 (32%). Therefore, the

Table 13. OLS regressions.

Source	**SS**	**df**	**MS**	Number of obs =		26
Model	1.92835336	3	.642784453	F (3, 22) =		3.09
Residual	4.57164664	22	.20780212	Prob > F =		0.0479
Total	6.5	25	.26	R-squared =		0.2967
				Adj R-squared =		0.2008
				Root MSW =		.45585
ydepvar	**Coef.**	**Std. Err.**	**t**	**P > |t|**	**[95% Conf. Interval]**	
logx2	.2870372	.1949793	1.47	0.155	–.1173251	.6913994
logx7	.3918386	.4223144	0.93	0.364	–.4839879	1.267665
logx10	–.1000314	.1437495	–0.70	0.494	–.3981496	.1980869
_cons	.4469003	1.215315	0.37	0.717	–2.073509	2. 96731

[14] Gujarati (2003).

[15] Note that the slope of a function can be greater than one, even if the values are all between zero and one.

Table 14. Probit regression.

Probit regression Log likelihood = –11.39635				Number of obs =		26
				LR chi2 (3)		13.25
				Prob > chi2		0.0041
				Pseudo R2		0.3676
ydepvar	**Coef.**	**Std. Err.**	**z**	**P > |t|**	**[95% Conf. Interval]**	
logx1	1.112119	.685899	1.62	0.105	–.232218	2.456457
logx7	4. 945958	2.509466	1. 97	0.049	.0274952	9.864421
logx10	–.8735025	.5370388	–1.63	0.104	–1.926079	.1790742
_cons	–6.138373	5.072515	–1.21	0.226	–16.08032	3.8035739

Table 15. Marginal effects after probit.

Marginal effects after probit y = Pr(ydepvar) (predict) = .34080523							
Variable	**dy/dx**	**Std. Err.**	**z**	**P > |z|**	**95% Conf. Interval**		**X**
logx1	.4078604	.24942	1.64	0.102	–.08099	.896711	–2.94363
logx7	1.813888	.81946	2.21	0.027	.207781	3.42	.760943
logx10	–.3203497	.18405	–1.71	0.082	–.681084	.040385	–5 99676

results support the three adduced hypotheses that the main root causes of the Darfur conflict are the land-tenure problem, ethnic tension about appointments of native administration (law and order), and desertification.

Concluding Remarks

It is clear that the hypothesized institutional failures of setting the land tenure system (property rights), controlling desertification, and adjudication of local law and order by native administration subjugated the people of Darfur to a conflict that has been hard to resolve. The focus of previous studies has mainly been on the land-tenure problem. This study tested the impact of several other variables, commonly perceived as root causes. The existence of their high multicollinearity with land tenure, desertification, and local law and order indicates that they are catalysts rather than initiating causes. Stepwise probit analysis confirms the validity of the three hypotheses. These results are consistent with the results of the survey done by the author in 2007–2008 and the conclusions of previous studies. It, therefore, behooves policy makers and other organizations seeking to find possible solutions for the Darfur conflict to negotiate a peaceful resolution to property rights, curbing desertification, and proper administration of setting the local law and order.

References

Abdul-Jalil. 2006. The dynamics of customary land tenure and natural resource management in Darfur. Land Reform, Settlement, and Cooperatives, 01.

Abdul-Jalil, Mjusa and Jon Unruh. 2013. Land rights under stress in Darfur: A volatile dynamic of the conflict. War and Society, 32(2), 156–181.

Agricultural Planning Unit. 2006. Agricultural Survey 2006/2007, Ministry of Agriculture, Animal Resources, and Irrigation, November.

Ahlerup, Pelle and Ola Olsson. 2012. The roots of ethnic diversity. Journal of Economic Growth, 17, 71–102.

Al Mangouri, Hassan Abdalla. 2004. Combating Desertification: Experience from Umm Kaddada District in East Darfur. University of Peace Conference Proceedings, December, Khartoum.

de Waal, Alex. 1989. Famine That Kills: Darfur, Sudan1984–1985, Oxford: Clarendon Press.

de Walle, Nicolas Van. 2001. African Economies and the Politics of Permanent Crisis, 1979–1999. Cambridge, UK: Cambridge University Press.

Department of Environment and Desertification in EL Fashir, Darfur. 2016.

Douglas Hibbs and Ola Olsson. 2004. Geography, biogeography and why some countries are rich and others are poor. Proceedings of the National Academy of Sciences of the USA, March 9, 101(10), 3715–3720.

Fadul and Abduljabbar Abdalla. 2004. Natural resources management for sustainable peace in Darfur. pp. 33–46. In University for Peace. Environmental Degradation as a Cause for Conflict in Darfur. Conference Proceedings, December.

Fadul, Abdul-Jabbar and Victor Tanner. 2007. Darfur after Abuja: A view from the ground. pp. 284–313. *In*: de Waal and Alex (eds.). War in Darfur and the Search for Peace. Cambridge, MA: Global Equity Initiative Harvard University.

Gujarati, Damodar N. 2003. Basic Econometrics, 4th ed., 580–633.

Ibn Khaldoun and Abdul Rahman Ibn Mohamed. 1377. Mugadimat Ibn Khaldoun, Dar Elgeel, Beiruit.

International Country Risk Guide. 2018. The PRS Group.

Kevane, Michael and Leslie Gray. 2008. Darfur: rainfall and conflict. Environmental Research Letters, 3, 1–10.

Mangouri, Hassan, Abdalla Al. 2004. Combating desertification: Experience from Umm Kaddada district in East Darfur. pp. 47–58. In University for Peace. Environmental Degradation as a Cause for Conflict in Darfur. Conference Proceedings, December.

Ministry of Agriculture. 2016. Department of Statistics, Sudan.

Mascona, Jacob, Nunn, N. and Robinson, J. A. 2017. Social Structure and conflict: evidence from Sub-Saharan Africa. NBER Working Papers, April, 1–64.

Olsson, Ola, Måns. Bruun and Henrik G. Smith. 2002. Starling foraging success in relation to agricultural land-use. Ecogeography, 25(3), 363–371.

Olsson, Ola and Heather Cogdon Fors. 2004. Congo: The prize of predation. Journal of Peace Research, 41(3), 321–336.

Olsson, Ola. 2010. After the Janjaweed? Socioeconomic impacts of the conflict in Darfur. The World Bank Economic Review, 24(3), 386–411.

Olsson, Ola and Eyerusalem Siba. 2013. Ethnic cleansing or resource struggle in darfur? an empirical analysis. Journal of Development Economics, 103, 299–312.

Rabbah, Nazik El Tayeb. 1998. Dawr El Hukuma El Markazia Wa El Idara El Ahlia Fi Fad El Niza Fi Darfur, MSC Thesis, Department of Political Science, Faculty of Economics and Social Studies, University of Khartoum.

Suliman, Osman. 2011. The Darfur Conflict: Geography or Institutions? Routledge.

Takana, Yousif. 2007. Mushkilat Darfur: Fashal Binaa El Dawla El Watnia. El-Ayam Daily, no. 8789, Thursday, April 26.
UNEP. 2007. Sudan Post-Conflict Environmental Assessment. Draft. January.
Unruh, Jon. 2012. Land and legality in Darfur conflict. African Security, 5, 105–128.
Unruh, Jon and Musa Abdul-Jalil. 2012. Land rights in Darfur: Institutional flexibility, policy, and adaptation to environmental change. Natural Resources Forum, 36, 274–284.
Unruh, Jon and Musa Abdul-Jalil. 2014. Constituencies of conflict and opportunity: land rights, narratives, and collective action in Darfur. Political Geography, 42, 104–116.
Vanrooyen, Michael, Jennifer Leaning, Kirsten Johnson, Karen Hirschfeld, David Tuller, Adam Levine and John Hefferman. 2008. Employment of a livelihoods analysis to define genocide in the Darfur region of Sudan. Journal of Genocide Research, 10(3), 343–358.
Young, Helen and Abdal Monim Osman. 2006. Challenges to peace and recovery in Darfur: a situation analysis of the ongoing conflict and its continuing impact on livelihoods. Feinstein International Center Briefing Paper, December.

CHAPTER 5

Presenting the Challenges and Offering New Strategies to Private and Public Organizations for Environmental Sustainability

*Hasan Volkan Oral** and *Hasan Saygın*

Introduction

Rapid population growth has brought some basic problems. The uncontrolled use of natural resources due to this increase has caused irreversible major environmental drawbacks. The most important of them are climate change, air water and land pollution, environmental degradation, industrialized agricultural practices and the reduction of natural resources. All of these environmental problems that have arisen have a chain effect on the formation of another. To find out some reliable solutions for these problems, the concept of 'Sustainable Development' was introduced in the early 1980s to create public opinion and policy. International Union for Conservation of Nature (IUCN 1980) published a report titled "World Conservation Strategy: Living Resource Conservation for Sustainable Development", which defined the principle of sustainable development. Seven years later, in 1987, another report published, which referred the IUCN's report "Our Common Future" or commonly known as "Brundlandt Report" by the UN's World Commission for Environment. "Our Common Report" is significant because the public awareness on this topic has increased significantly and noteworthy after publishing the report (Oral 2020). Added to that this report, a comprehensive one produced through a global partnership constituted a major political turning point for the concept of sustainable development (Mebratu 1996).

İstanbul Aydın University, Faculty of Engineering, Florya Halit Aydin Campus, K. Cekmece Istanbul, Turkey.
Email: hasansaygin@aydin.edu.tr
* Corresponding author: volkanoral@aydin.edu.tr

Although the three pillars of sustainability, environmental, social and economic are presumed to work in harmony, in the real world there are often conflicts among the three. For example, managers of companies wonder whether it pays to be green or not (Jayanti and Gowda 2014). The concept of environmental sustainability includes social sciences, as well as engineering sciences, offer solutions for these problems. Within this context, the most important approach is the ability of the public and private organizations to act in coordination, which includes planning, organization and efficient human resource management terms.

Public organizations serve a government, apparently in the public interest and a wide range of private organizations have interests in the research or policy aspects of health, and some are specifically focused on the issues around developing evidence-informed policy. Non-governmental organizations (NGOs) that support studies and activities to help provide evidence-informed health policy, consulting companies that engage in the contract as well as in independent research in policy areas, and foundations and professional societies that support policy-related research as a primary function can be given as an example (Pomeroy and Sanfilippo 2015). Environmental sustainability practices in the business world have been previously studied by Gerbens-Leenes et al. (2003), Seow et al. (2006), Al Khidir and Zailani (2009), Roxas and Coetzer (2012), Henderson (2015), Bocken et al. (2014), Rajala et al. (2016), Bamgebade (2019), and Rajeh (2020) in the literature. Gerbens-Leenes et al. (2003) studied sustainable company performance and proposed the development of a practical measuring system in firms. Seow et al. (2006) offered to introduce a conceptual methodology to support decisions about environmental systems. They suggested that potential managerial applications and implications include fields, for instance, product cost management, business process design and technology selection could be applied in an integrated way. Al Khidir and Zailani (2009) made the definitions related to the green supply chain, which is so important for the business environment in Malaysia. Roxas and Coetzer (2012) have examined the direct impact of three dimensions of the institutional environment on managerial attitudes toward the natural environment and the direct influence of the latter on the environmental sustainability orientation (ESO) of small firms. Henderson (2015) investigated the answer to the question as "Can a business case be made for acting sustainably?" and according to obtained results this could be achieved, but the first step is relevant to find out the generally accepted definition of "sustainability". Bocken et al. (2014) prepared a review study to develop sustainable business model archetypes in companies and they concluded that the literature and practice of sustainability innovation in this field are vast but fragmented. Rajala et al. (2016) suggested that business model greening in companies requires a major change in the company's ecosystem and this model is a multi-layered process rather than a phased model. Bamgebade (2019) examined the factors which are driving ecological sustainability in construction firms. They reported that firm capabilities are crucial to improve environmental sustainability in large firms. Rajeh (2020) studied the sustainability performances of 39 firms in the Indian context listed in Environmental, Social and Governance (ESG). According to their findings, resource use, environmental innovation score and company social responsibility (CSR) approach scores emerge to be as important. When these studies are analyzed in detail, it is observed that the adaptations of businesses to environmental sustainability are examined from different perspectives. The most striking point in these studies is making the definition of sustainability and how this concept should be adapted to the business management culture. However, innovative perspectives emerging with today's developing technology and suggestions such as understanding existing problems and offering integrated solutions related to these problems are not mentioned or investigated. To evaluate in a detailed way related to sustainable strengths, weaknesses, opportunities and threats (SWOT) analyzes, green human resources management and digital infrastructure applications were never mentioned. Why was this book chapter typed? Company managements are still not fully aware of how Sustainable Development Goals (SDGs) should be implemented in practice. It is a very challenging process to continue this most efficiently without reducing production and to fulfill the requirements of environmental sustainability at the same time. This book chapter is typed to identify the current problems in the implementation of environmental sustainability for public and private organizations. The further step is also to offer innovative solutions and strategies to overcome these problems. The solutions have been offered by taking into consideration the technological innovations of today.

Therefore, the main purpose of this chapter is to define and examine the problems related to environmental sustainability relevant to environment and development context in both public and private organizations and to provide new strategies and policy implications. The challenges between environmental sustainability and development in public and private organizations can be listed as follow:

- Failure to perform (SWOT) analysis or a similar analysis properly to reveal the strengths and weaknesses of institutions.
- Failure to make business plans and to define accurate job descriptions according to the capabilities of staff.

The organizational structure of this chapter (Figure 1) is prepared due to these challenges. Section 1 'Problem Definition 1' presents the SWOT analyzes about the weaknesses of institutions. Section 2 'Problem Definition 2' is about making the business plans and highlighting the importance of defining accurate job descriptions according to the capabilities of staff in the organizations, and Section 3 'New Solutions and Strategies' offer novel strategies and suggest strategic and policy implications. Sections 4 and 5 are labeled as 'Discussions and Conclusions' regarding offered strategies, respectively.

Problem Definition 1

- Strengths, Weaknesses, Opportunities and Threats (SWOT) Analysis in the Public and Private Organizations Related to Sustainability

  There is no unique definition of sustainability for organizations (Pojasek 2012). The definition made by the Brundtland Report does not present the operational way of application of sustainability in the organizations. As the result, Pojasek (2012) made the following description of the sustainability, which clarifies the understanding related to operational behavior in the organizations: "Sustainability

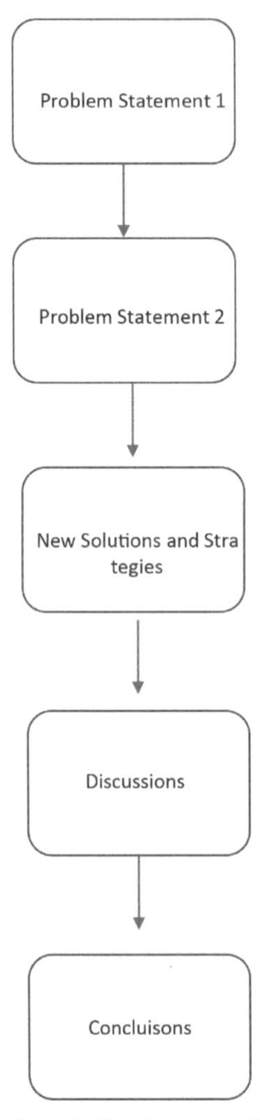

Figure 1. Organizational structure of the study.

is the capability of an organization to transparently manage its responsibilities for environmental stewardship, social well-being, and economic prosperity over the long term while being held accountable to its stakeholders". Based on this definition, the most efficient analysis technique for the sustainability concept to be applied in organizations is the SWOT analysis. SWOT stands for strengths, weaknesses, opportunities and threats. SWOT analysis, which was developed in the 1960s, is a strategic assessment tool for gathering and organizing the information needed to evaluate positive (strengths and opportunities) and negative (weaknesses and threats) elements of a strategy, project, business model, company or industry, and these analyzes can be easily implemented in sustainability in public and private organizations using the International Standards Organizations (ISO) 14001 criteria (Pesce et al. 2018). Briefly, ISO 14001 is a set of standards that cover environmental management systems (EMS). EMS is intended to formalize procedures for managing and reducing environmental impacts in the companies (Christini et al. 2004). According to Pesce et al. (2018), sustainability-related aspects of ISO 14001 that affect firms can be grouped under the elements of SWOT analysis. As for strengths, the idea of sustainability promotes the idea of environmental protection and environmental awareness in a company and it guides how to improve the environmental sustainability of a company. As for opportunities, more public attention will be provided for environmental protection, making a product more environmentally friendly, which is also known as eco-friendly production, and improving the environmental communication of companies can be counted as the benefits. On the contrary, mentioning life cycle thinking but does not require life cycle assessment (LCA) and not legally binding can be given as the examples for the weaknesses. Moreover, obtaining environmental certifications which are outcomes from external pressure, not self-driven and having not to aim to protect the environment in the companies are the other examples related to threats. According to Bull et al. (2014), the value of a SWOT analysis stems not only from its ability to highlight ways in which an agent's internal and external environments interact to affect its success but also from its ability to be used in the development and implementation of long-term strategies to achieve particular objectives. Bull et al. (2014) also mentioned that there are various classes of strategies that can follow from a SWOT analysis, e.g., those that link Strengths and Opportunities ('SO Strategies'), those that link Weaknesses and Opportunities ('WO Strategies'), those that jointly focus on the Strengths and Threats ('ST strategies') and those that arise from the joint assessment of Weaknesses and Threats ('WT Strategies'). For instance, SO strategies utilize the fact that 'strengths' may help to capitalize upon external 'opportunities', whereas WO strategies focus upon the pursuit of external 'opportunities' to lessen the severity of 'weaknesses'. Similarly, ST strategies focus on the potential for existing internal 'strengths' to mitigate the impact of external 'threats', while WT strategies consist of actions intended to reduce both internal 'weaknesses' and external 'threats' simultaneously. Guta (2015) defined planning, organization, coordination, drive and control-evaluation are the most problematic functions in an organization. At the planning stage, during the preparation of medium- and long-term strategies, it is extremely low, the share of current activities being superior; deadlines are generally very restricted activities are concentrated mainly on results than on process. At the organization stage, unscheduled share current activities planned is superior effect being noticed by the difficulty which generates various projects of local interest; relationships between departments held strictly limit the need for registering major deficiencies in the communication; resources allocated are insufficient relative to current needs; the procedures and standards are mostly bureaucratic, characterized by an extensive informational circuit, which affects the time necessary decisions. As for coordination, the process of recruitment and selection is conducting as a formality, and many posts were occupied before the launch of the necessary formalities; in terms of personnel training, there is a training plan specific to the institution based on actual training needs. For drive and control-evaluation steps, low importance is given to this function of management; lack of a system to monitor the degree of motivation which permits identification of grievances and improving the current situation by integrating the positive and mitigate the negative effects and weight control activity is superior to the evaluation; lack of monitoring activity in the evaluation process; there are no criteria for performance evaluation.

Problem Definition 2

- Making Business Plans and Defining Accurate Job Descriptions According to the Capabilities of Staff in Public and Private Organizations

  In any type of business, to be able to conduct manufacturing with reduced time and energy is a preferable situation, but this cannot be taken place as desired. Producing something more with less of anything is also a way of ecological footprinting (Thorpe 2019). As a rule of thumb, energy is used as an indispensable resource for obtaining a product, both at public or private organizations. The efficient use of this resource is a method that will put the private and public sectors advantageously in terms of manufacturing (Ertaş 2017). The efficient use of energy in every organization is through the implementation of the right and accurate business plans. A business plan is an effective tool used by organizations to schedule the goals and objectives in a coherent format. It can be defined as operating a company on paper. No matter the size or stage of development, companies use business plans to improve internal operations and to describe and market the business to potential outside financiers (Hormozi et al. 2002). Honig and Karlsson (2004) pointed out that writing a business plan is an institutional behavior and survival and profitability are the well-known outcomes of these plans. When a plan is designed human, social capital and demographic variables of the organizations are needed to be careful of. As for the outcomes, the relationship between business planning and survival of the nascent organization, the founders have a business education that has a greater propensity to produce business plans than those whose founders do not have a business education has implications for arguments based both on economic rationality and institutional theory. From an economic point of view, survival represents the success of the organization. From an institutional perspective, completing a business plan yields the nascent entrepreneur increased legitimacy, even in the event of sustained economic losses. For profitability, business plans are based on the premise of the rational economic actor. From this perspective, the business plan is a rational activity that assists the owners of new firms (entrepreneurs) to earn larger profits through efficiency gains and/or increased sales. Regardless of the type of organization, the most basic mistakes seen in a business plan are making short and medium-term strategic plans. However, these plans should be made in the long term in the tactical infrastructure (Fry and Stoner 1985). Undoubtedly, making business plans alone does not make sense in organization management. Business plans should be supported by well-planned and high-quality job descriptions designed according to the working purpose of the organizations. Rusak (2018) emphasizes that effective well-written job descriptions have the following benefits: candidate attraction, hierarchy, performance management, new employee orientation, training and development and the basis for compensation. Consequently, it can be said that the high performance of an organization is depending on the capacities of the employees. This term is also linked to another term, which is known as capacity building. According to Yamoah and Maiyo (2013) capacity building has different meanings and interpretations depending on who uses it and in what context it is used. It is generally accepted that capacity building as a concept is closely related to education, training and human resource development in an organization. This conventional concept has changed in recent years toward a broader and more holistic view, covering both institutional and country-specific initiatives. Yamoah and Maiyo (2013) also stated that capacity building as the development of knowledge, skills and attitudes in individuals and groups of people relevant in design, development, management and maintenance of institutional and operational infrastructures and processes that are locally meaningful. As a result, the capacities of the employees or the employee performance both in public and private organizations are about to review employee performance against standards set and identify strengths and weaknesses of individuals, both in terms of personal characteristics and delivering skills (Yamoah and Maiyo 2013). According to a report published by the Australian Public Service Commission Report (2018) to achieve high performance, organizations must understand how their resources, routines, structures, systems and processes are brought together and leveraged to support high performance. They must also understand how these organization-wide mechanisms enable them to adapt and change. The focus here is on ensuring that the skills, knowledge, routines and processes evident within organizations complement one another and enable high performance. It is important to note that, in most cases, the strategic leveraging of capabilities requires recognition and operationalization of existing internal strengths rather than building new ones; the key challenge being how to identify, encourage and manage them, highlighting the importance of leaders in enabling and managing capabilities.

New Solutions and Strategies

Until this section, the problems encountered in the environment and development axis in the public and private sectors were tried to be examined under certain headings. In this section, strategies, strategic and policy implications will be discussed for the problems encountered.

a. *Offered New Strategies for Section 1*

- **Conducting modified SWOT analyzes in private/public organizations due to environmental sustainability assessments:** SWOT analyzes coincide with the 1950s and early 1960s. The discussion of the concept of sustainability in the scientific literature coincides with the early 1980s. As a result, SWOT analyzes should be reorganized according to sustainability parameters and applied in companies. The new SWOT analysis to be applied can be called 'swot' by taking the 'S' of sustainability. The Sustainability SWOT Analysis (swot) (Pesonen and Horn 2003, Azapagic 2003, Metzger et al. 2012) helps companies assess environmental risks and to drive action on environmental challenges. It helps individuals to engage and motivate colleagues. Along with the conventionally applied SWOT analysis parameters, a new evaluation criterion can be added under environmental challenges. The criteria to be added can be gathered under the environmental sub-topics as natural challenges, scarcity, water availability, waste and hazards, global warming and climate variability and extreme as environmental challenges (Tools and for Sustainable Product and Design 2020). Metzger et al. (2012) reconsidered the existing SWOT analysis with a new structure based on the triangle hierarchy, including the environmental parameters. In order for this newly established structure to be more understandable, the following questions were also asked at each stage to be followed in the hierarchy; for environmental challenges and big trends step "what or (who) do you want to inform?" and "what do you and other see changing?", for the threats step "where are environmental challenges creating broad threats to future business value", for the opportunities step "where is there a growing gap where we and others can create new solutions for environmental challenges?", for the strengths step "what are unexpected ways we can apply our strengths to environmental challenges?", for the weakness step "who else has similar weaknesses or faces similar risks from environmental challenges?", for the prioritize step "which insights will influence or keep/him/her up at night?" and for the last step known as the act, "what can we do in the near-term, mid-term and long-term?".

- **Conducting a multi-criteria decision-making based method (MCDM) SWOT analyzes:** The MCDM can be used to support SWOT analyses. The main purpose of MCDM is to rank different models approximately. As a result, MCDM offers a very useful structure for SWOT analyzes and could be used in a semi-quantitative matter (De Wilde et al. 2017). For instance, the integration of these two modeling approaches can be seen at a European Union 7th Framework Project, 4 Fun, which stands for 'The FUture of FUlly integrated human exposure assessment of chemicals'. This project aims at delivering a standardized tool for human exposure assessment to chemicals, named 'MERLIN-Expo'. The MERLIN-Expo tool integrates on the same platform multimedia Physiologically based pharmacokinetic (PBPK) and dose-response models allowing to cover all the exposure assessment chain (from concentration in water, air and/or soil to internal dose to target organs and eventually pathology risks). In this way, it will be possible to carry out lifetime risk assessments for different human populations, including exposure through multiple pathways (EUP 2015). The theoretical background of this model can be applied to determine the risk assessment studies at private and public companies, respectively.
- **Launching a life cycle sustainability assessment-based decision-analysis framework:** As previously explained in Section 1, LCA stands for Life Cycle Assessment and life cycle sustainability assessment (LCSA) can also be used with the term 'sustainability'. LCSA to inform and guide decision-makers through a step-by-step development of sustainability solutions for production systems. LCSA provides a systematic and life cycle evaluation of the sustainability

impacts in product systems. The LCSA driven solutions are developed based on a systematic and explicit decision-analysis approach that assists in the understanding of the problems on the three dimensions of sustainability, generation of solutions, evaluation, acceptability and manageability tests and making recommendations (Hannouf and Assefa 2018). For instance, this system can be applied to water and energy consumption parameters consumed by both private and public organizations. Thus, organization managers can implement saving practices for whichever of these parameters is used extremely.

b. Offered New Strategies for Section 2

- **Digital-based integrated sustainability reporting:** Maintaining organizational management with digital infrastructure is one of the concepts that we have encountered in the last five years (Legner et al. 2017, Coreynen et al. 2017, Rachninger et al. 2019). In a broad sense, this process can also be called as 'Digital Business Transformation'. It is the application of technology to build new business models, processes, software and systems that result in more profitable revenue, greater competitive advantage and higher efficiency. Businesses achieve this by transforming processes and business models, empowering workforce efficiency and innovation and personalizing customer/ citizen experiences. Transforming the business model is done through digital business modification, new digital business and digital globalization. These processes take place by adding digital content to existing products and services and introducing new digital solutions (Shvertner 2017). Sustainability reports are a type of report published by many company management. With the introduction of the concept of sustainability, integrated sustainability reports (Bini and Belluci 2020) have been published by company management in recent years. At this point, the greatest success for organizations might be to ensure directly integrating sustainability reporting into management practices (Adams and Frost 2008). After this step is launched, applying the concepts of combining integrated sustainability reporting and performance control (Seele 2016) can yield efficient results for company management.
- **Green Human Resources Department Management (HRDM):** According to Schroeder (2012), HRDM is an important concept for strategic sustainability in the organizations and this term closely is associated with significant business benefits as well as positive environmental impacts, yet many organizations fail to recognize the potential of this approach and neglect the factors necessary for its successful implementation. Incorporating specific sustainability-related goals into job descriptions in the organizations and core competencies will also help to ensure that employees are motivated to work toward sustainability goals and that these receive similar levels of attention as any other criteria against which individuals and teams are formally assessed. Moreover, high-quality HR policies and practices integrated with business and sustainability goals are likely to promote positive employee-related outcomes, such as improved morale increased engagement and higher productivity. This has an impact on improved retention and improves the company's brand image which in turn is likely to

increase sales and increase the organization's attractiveness to potential recruits. A new approach has been included in the HR concept for the past few years. This application is known as Green HRM (Yong et al. 2019, Mousa and Othman 2020, Singh et al. 2020). This term is still being discussed in the literature but in general, it can be defined as practices that can help the organizations align their business strategies with the environment (Yongh et al. 2019). Added to that, Green HRM refers to HRM practices aimed at the environmental and ecological influence of the organizations, and it is linked with firm environmental strategy and green behaviors of employees (Singh et al. 2020). Moreover, the organizations' management has to pay attention to working with the skilled HR Department. For instance, the HR Department Head or the staff needs to know the meaning of environmental sustainability or have an idea about the importance of Sustainable Development Goals at least.

- **Improving employees' capacities related to the sustainable way of business:** The achievement in the development of an organization also depends on the capacities of the employees. Organizations can adapt and launch various HRM practices to enhance and improve employee skills. First, efforts are on improving the quality of the individuals hired or on raising the skills and abilities of current employees or both. This can be achieved by providing comprehensive training on the job and development activities after selection (Wanyama and Mutsotso 2010). While these activities are being carried out, some concepts such as the sustainable way of business perspective and environmental sensitivity can be taught to the employees with training programs, which can be provided by the experts of the subject.

c. *Strategic and Policy Implications*

The public and private organizations need to provide statements to find some solutions for the challenges between environmental sustainability and development. First of all, the organization managers should undertake that they will fulfill all the practices of the environmental management system, which includes the latest updated standards. This commitment should be adopted as company policy. Organizations should indicate to employees that it is a company policy to demonstrate environmentally sensitive behavior in their employment contracts during the recruitment. It should be noted that employees who do not act according to environmental sustainability rules can be dismissed or the job contracts will be terminated. Routine environmental sustainability training of all employees should be announced to the employees by the organization management as an organization rule.

Discussion

Although environmental sustainability practices in the business world are handled by various researchers (Gerbens-Leenes et al. 2003, Seow et al. 2006, Al Khidir and Zailani 2009, Roxas and Coetzer 2012, Henderson 2015, Bocken et al. 2014, Rajala et al. 2016, Bamgebade 2019, Rajeh 2020), the challenges that are presented in this book and their solution suggestions have not been investigated by these researchers, respectively. Added to that the concept of sustainability has been added to the SWOT analysis were studied by Pesonen and Horn (2013), Azapagic (2003) and Metzger et al. (2012), but the feasibility and applicability of this assessment by the organization managements are only suggested in this book chapter. Besides that, it is inevitable to monitor sustainable SWOT analysis, which is based on MCDM approaches, that can provide great benefits to company management in terms of implementing environmental sustainability.

The LCSA driven solutions recommended by Hannouf and Asefa (2018) could be another useful method that allows practicing environmental sustainability in the organizations. Green human resources management (Yongh et al. 2019, Mousa and Othman 2020, Singh et al. 2020) brings a different perspective to classical human resources management which should be acknowledged as an application and should be included in the business strategy, management culture and philosophy of the organizations in the coming years. Based on this new human resource perspective and idea, organization management can have a chance to implement the actions that will increase their employees' capacities based on a sustainable way of thinking. It should not be overlooked that the digitization applications, which have emerged with the rapid development of information and internet technology in recent years, offer wide opportunities to company management in terms of the implementation of environmental sustainability.

Conclusions

The most important contribution of this study to company management is that it is a guide on the problems that companies may encounter during environmental sustainability practices. Solution and strategy suggestions were made by explaining what these problems might be. For the strategies to be developed to solve these problems, firstly it is important to raise the perception and awareness levels of the employees working in the organizations primarily on environmental issues. Secondly, related to organization management, the process of digitization should be followed by developing qualified business plans that need to integrate. Finally, the strategies which will be a solution for the detected problems can be listed as follows: conducting modified SWOT analyzes in private/public organizations due to environmental sustainability assessments, conducting multi-criteria decision making based method (MCDM) SWOT analyzes, launching a life cycle sustainability assessment-based decision-analysis framework, digital-based integrated sustainability reporting, establishing green human resources department management and improving employees' capacities related to a sustainable way of business.

References

Al Khidir, T. and Zailani, S. 2009. Going green in supply chain towards environmental sustainability. Global Journal of Environmental Research, 3(3), 246–251.

Australian Public Service Commission Report. 2018. Annex 9: Capabilities: competencies of individual employees, competences and dynamic capabilities of organizations. https://www.apsc.gov.au/. Access Date 26.02.2020.

Azapagic, A. 2003. Systems approach company sustainability: a general management framework. Trans IChemE, 81(5).

Bamgebade, J. A. 2019. Analysis of some factors driving ecological sustainability in construction firms. Journal of Cleaner Production, 208, 1537–1545.

Bini, L. and Belluci, M. 2020. Integrated Sustainability Reporting. Springer Publications, Switzerland.

Bocken, N. M. P., Short, S. W., Rana, P. and Evans, S. 2014. A literature and practice review to develop a sustainable business model archetypes. Journal of Cleaner Production.

Bull, J. W., Jobstvogt, N., Bohnke-Henrichs, A., Mascarenhas, A., Sitas, N., Baulcomb, C., Lambini, C. K., Rawlins, M., Baral, H., Zahringer, J., Carter-Silk, E., Balzan, M. V., Kenter, J. O., Hayha, T., Petz, K. and Koss, R. 2014. Strengths, weaknesses, opportunities, and threats: A SWOT analysis of the ecosystem services framework. Ecosystem Services, 17, 99–111.

Carol, A. Adams and Frost, G. R. 2008. Integrating sustainability reporting into management practices. Accounting Forum, 32(4), 288–302, DOI: 10.1016/j.accfor.2008.05.002.

Coreynen, W., Matthyssens, P. and Van Bockhaven, W. 2017. Boosting servitization through digitization: Pathways and dynamic resource configurations for manufacturers. Industrial Marketing Management, 60, 42–53. https://doi.org/10.1016/j.indmarman.2016.04.012.

Christini, G. 2004. Environmental management systems and ISO 14001 certification for construction firms. Journal of Construction Engineering and Management, 130(3), 330.

De Wilde, T., Verdonck, F., Tediosi, A., Tanaka, T., Bonnard, R., Banjac, Z., Isigonis, P., Giubilato, E., Critto, A., Zabeo, A., Suciu, N., Garrat, J. and Ciffroy, P. 2017. SWOT analysis of the MERLIN-Expo and its relevance in legislative frameworks. *In*: Philippe Ciffroy, Alice Teodosi and Ettore Capri (eds.). Modeling the Fate of Chemicals in the Environment and Human Body. Springer International Publishing, Cham, Switzerland.

Ertaş, H. 2017. Performance management in the public sector as motivation and success factor: a theoretical analysis. Kastamonu Üniversitesi İktisadi ve İdari Bilimler Fakültesi Dergisi.

European Union Project (EUP). http://4funproject.eu/ Access Date 26.02.2020.

Fry, F. L. and Stoner, C. L. 1985. Business plans: Two major types. Journal of Small Business Management, 23(1).

Gerbens-Leenes, P. W., Moll, H. C. and Schoot Uiterkamp, A. J. M. 2003. Design and development of a measuring method for environmental sustainability in food production systems. Ecological Economics.

Guta, A. J. 2015. Analysis of the public management administration in Romanina based on diagnostics and SWOT analysis. Annals of the University of Petroşani, Economics, 15(1), 147–154.

Hannouf, M. and Assefa, G. 2018. A life-cycle sustainability assessment-based decision-analysis framework. Sustainability, 3863; https://doi:10.3390/su10113863.

Henderson, R. 2015. Making the business case for environmental sustainability. Harvard Business School Working Paper, No. 15-068.

Honig, B. and Karlsson, T. 2004. Institutional forces and the written business plan. Journal of Management, 30(1), 29–48.

Hormozi, A., Sutton, G., McMinn, R. and Lucio, W. 2002. Business plans for new or small businesses: paving the path to success. Management Decision, 40(8), 755–763. https://doi.org/10.1108/00251740210437725.

International Union for Conservation of Nature (IUCN). 1980. World Conservation Strategy: Living Resource Conservation for Sustainable Development. https://portals.iucn.org/library/efiles/documents/WCS-004.pdf.

Jayanti, R. K. and Gowda, M. V. R. 2014. Sustainability dilemmas in emerging economies. IIMB Management Review, 26(2), 130–142.

Laura, Bini and Bellucci, Marco. 2020. Integrated Sustainability Reporting: Linking Environmental and Social Information to Value Creation Processes. 10.1007/978-3-030-24954-0.

Legner, C., Eymann, T., Hess, T., Matt, C., Bohman, T., Drews, P., Madche, P., Urbach, N. and Ahlemann, F. 2017. Digitalization: opportunity and challenge for the business and information systems engineering community. Business & Information Systems Engineering, 59, 301–308. https://doi.org/10.1007/s12599-017-0484-2.

Mebratu, D. 1996. Sustainability and sustainable development: historical and conceptual review. Environmental Impact Assessment Review, 18(6), 493–520, https://doi.org/10.1016/S0195-9255(98)00019-5 (Access Date: 10.05.2018).

Metzger, E., Putt del Pino, S., Prowitt, S., Goodward, J. and Perera, A. 2012. A Sustainability SWOT. World Resources Institute.

Mousa, S. K. and Othman, M. 2020. The impact of green human resource management practices on sustainable performance in healthcare organizations: A conceptual framework. Journal of Cleaner Production, 243, 1158595.

Oral, H. V. 2020. Sustainable development. *In*: Scott Nicholas Romainuk, Manish Tapa and Peter Marton (eds.). The Palgrave Encyclopedia of Global Security Studies. Palgrave Springer Publications, N.Y.

Pesce, M., Shi, C., Critto, A., Wang, X. and Marcomini, A. 2018. SWOT analysis of the application of international standard ISO 14001 in the Chinese context. A case study of Guangdong province. Sustainability, 10(9), 3196.

Pesonen, H. and Horn, S. 2013. Evaluating the sustainability SWOT as a streamlined tool for life cycle sustainability assessment. International Journal Life Cycle Assessment, 18, 1780–1792. https://doi.org/10.1007/s11367-012-0456-1.

Pojasek, R. B. 2012. Understanding sustainability: an organizational perspective. Environmental Quality Management, 93–100. http://doi.10.1002/tqem.

Pomeroy, C. and Sanfilippo, F. 2015. How research can and should inform public policy. pp. 179–191. *In*: Steven Wartman (ed.). The Transformation of Academic Health Centers: The Institutional Challenge to Improve Health and Well-Being in Healthcare's Changing Landscape. Academic Press, U.S.A.

Rajala, R., Westerlund, M. and Lampikoski, T. 2016. Environmental sustainability in industrial manufacturing: Re-examining the greening of Interface's business model. Journal of Cleaner Production.

Rajeh, R. 2020. Exploring the sustainability performances of firms using environmental, social, and governance scores. Journal of Cleaner Production, (247), 119600.

Rachinger, M., Rauter, R., Müller, C., Vorraber, W. and Schirgi, E. 2019. Digitalization and its influence on business model innovation. Journal of Manufacturing Technology Management, 30(8), 1143–1160.

Roxas, B. and Coetzer, A. 2012. Institutional environment, managerial attitudes and environmental sustainability orientation of small firms. Journal of Business Ethics, 111, 461–476.

Rusak, S. 2018. The importance of having well-written job descriptions. https://cpl.com/ (Access Date: 26.02.2019).

Seele, P. 2016. Digitally unified reporting: how XBRL-based real-time transparency helps in combining integrated sustainability reporting and performance control. Journal of Cleaner Production, 136(Part A), 65–77.

Seow, C., Hillary, R., Sarkis, J., Meade, L. and Presley, L. 2006. An activity-based management methodology for evaluating business processes for environmental sustainability. Business Process Management Journal, 12(6), 751–769.

Schroeder, Harold. 2012. The importance of human resource management in strategic sustainability: an art and science perspective. Journal of Environmental Sustainability, 2(2), Article 4. DOI: 10.14448/jes.02.0004.

Shvertner, K. 2017. Digital transformation of business. Trakia Journal of Science, 15, 388–393. 10.15547/tjs.2017.s.01.065.

Singh, S. K., Del Giudice, M., Chierici, R. and Graziano, D. 2020. Green innovation and environmental performance: The role of green transformational leadership and green human resource management. Technological Forecasting and Social Change, 150, 119762.

Thorpe, D. 2019. "One Planet" Cities: Sustaining Humanity within Planetary Limits. Routledge Publications N.Y.

Tools and for Sustainable Product and Design. 2020. https://www.threebility.com/. Access Date:26.02.2020.

Velasquez, M. and Hester, P. T. 2013. An analysis of multi-criteria decision making methods. International Journal of Operations Research, 10(2), 56–66.

Wanyama, W. K. and Mutsotso, S. 2010. Relationship between capacity building and employee productivity on the performance of commercial banks in Kenya. African Journal of History and Culture, 2(5), 73–78.

Yamoah, E. E. and Maiyo, P. 2013. Capacity building and employee performance. Canadian Social Science, 9(3), 42–45.

Yong, J. Y., Yusliza, M.-Y. and Fawehinmi, O. 2019. Green human resource management: A systematic literature review from 2007 to 2019. Benchmarking: An International Journal, https://doi.org/10.1108/BIJ-12-2018-0438.

Chapter 6

Sustainability from the Business Model to the Communities

Ciro Martínez Oropesa

Introduction

With the vertiginous increase of the commercialized goods trying to respond to unbridled market and consumer demand and consequently the negative impact of industrial activity on the planet forces us to consider the search for a new balance that allows harmonizing commercial interests with social development and environmental protection.

These challenges pose great challenges to the current industrial systems, the infrastructure at the service of production, distribution and consumption, which must go from being an essential part of the problem to being an active part of an essential transformation process in the use of resources and energy.

The objective of this work is to contribute to the definition of methods and practices that can improve the effectiveness of the sustainability management system as well as its close relationship with the organizational culture and the leadership of supervisors.

Sustainability cannot be a matter outside the industry; this is also demanded to promote best practices that seek to encourage the introduction of the sustainable development goals in its strategic planning by defining priorities, goals, implementation and reporting of information and applying principles of sustainability in its operation, which produce innovations for sustainable development.

Colombia, like most countries in the world, has been experiencing significant climatic changes with wide and sensitive impacts on its economy. On this subject, the (SDSN, Action Agenda Report 2014) recognizes that human societies are causing more harm than ever to the planet, and the environment inequality gaps between countries are increasing and governance is becoming more complex, especially with the increasing influence of technology.

Departamento Operaciones y Sistemas, Universidad Autónoma de Occidente, Cali, Colombia.
Email: cmartinezo@uao.edu.co

According to the Global Sustainable Development Report (2019), since the mid-twentieth century, human-made global change has accelerated sharply, creating multiple ecological pressures on Earth, pressures that are already too great to guarantee a safe habitat for future generations.

When it relates to the SDGs to the industry, there is a need for indicators, targets and early alerts that promote responsible consumption levels and determine production levels socially and responsibly that will minimize waste, the amount of energy and water associated with production and waste management.

This study includes successful methods and practices to improve the management systems implemented as well as the role that culture should play to consolidate the changes that the organizational situation demands.

The research question focuses on how organizational culture and leadership can influence the improvement of the effectiveness of the sustainability management system in organizations.

To achieve the above, the organizational culture and its relationship on the leadership (Block 2003) as their ability to influence to reach a long-term vision, supported by a solid management system, must mobilize all employees to project themselves in the triple value that constitutes sustainability—economic, social and environmental—while focusing on the organizations and communities.

In the work we offer, the author has the challenge to examine the question like what are the bases of the corporate sustainability management systems? It is known that this question has been trying to give the bests answer a few years ago, initially through the ISO 26000 standard that was developed before the 2030 agenda from the ONU and the ODS. Nevertheless, it is important to emphasize the aspect of leadership and organizational culture is essential to manage the changes demanded by the market and consumption as well as the effective management of priorities in the organization

To achieve the above, leadership, culture, management systems and the resources are fundamental pillars, and their ability to influence and mobilize toward a long-term vision. Senior management, supported by a solid management system, must mobilize all employees to project on the triple value that constitutes sustainability—economic, social and the environmental—while focusing on the interior of the organization and communities.

Initially, it is will briefly examine the theoretical thesis on organizational culture, its characteristics, functions and fundamental types, which will work as an introduction to understanding the relationship with the fundamental purposes and scopes of corporate sustainability management by reaching a process of cultural transformation that will allow aligning with new goals and demands of change that the organization must undertake. Next, the foundations of the management system are analyzed with greater emphasis on leadership, strategies and indicators that will establish the foundations of the process and the engine that will drive progressive changes.

Sustainability and Organizational Culture

Today's business world is immersed in constant and very rapid processes of change (the globalization of the economy, the rapid growth of electronic commerce, the growing pace of commercial operations, the rapid obsolescence of technological innovations and the rapid expansion of new companies in the world market) that inevitably imposes the need to develop new models and forms of leadership.

These chaotic and rapid changes in the environment create the risk that the organizational culture of the past inhibits the development of organizations rather than contributing to their corporate success. In accordance with by Kroeber and Parsons (1958), culture is defined as transmitted and created content and patterns of values, ideas and other symbolic-meaningful systems as factors in the shaping of human behavior. The organizational culture manifested in the shared assumptions, values, behavioral norms and practices, that characterize an organization, affects the way organization members interpret aspects of their work environment and create meaning in any given situation at work (Erdogan et al. 2006).

Culture is made up of different elements such as values, habits, uses and customs, codes of conduct, labor policies, traditions and objectives, which are internalized by individuals during the training process through education and socialization and are transmitted from generation to generation (Hall 1959).

In Colombia, studies carried out by different researchers show that organizational culture is essential in the study of organizations, regarding the creation of a management style, based on knowledge of the country's cultural context and its development strategies, taking into account the cultural problem as a basis for achieving productivity and quality in an integral way (Loaiza et al. 2019). Likewise, it is a topic of interest to study organization from different perspectives, leading to an analysis of the organization's context (Sánchez et al. 2006).

As it is observed, over time organizational culture's topic has gained prominence and importance in the field of organizational management; it is likely that previously its relevance was not noticed due to the lack of knowledge in the subject or the little analysis on the impact on the effectiveness in the organization since intangible results, which are difficult to treat, can be obtained (Calderón et al. 2010).

Organizational Culture

According to Tylor (1891), culture is considered as the integration of knowledge, belief, art, morals, law, custom, abilities and habits that each individual, who is part of society, possesses.

From the etymological point of view, culture refers to the Latin verb 'colo', 'colere', 'cultum' which expresses cultivation and is related to the word 'agriculture' interpreted as field cultivation. Therefore, culture from the etymological definition is education, training, development or improvement of man's intellectual and moral faculties (Altieri 2001). In accordance with the previous definitions, the Royal Spanish Academy (2019) states that culture is defined as the set of ways of life and customs, knowledge and degree of artistic, scientific, industrial development, at a time, social group, etc.

According to Belalcázar (2012), culture is a set of values that make up the organization, where norms, behavior patterns, artifacts, values, beliefs, thoughts, policies, etc., are involved with which the members of the organization feel more or less identified.

Other authors have defined it as those patterns of values, ideas and other systems focused on behavior (Kroeber and Kluckhohn 1952) or as a system of values and beliefs, which are related to people, the organization and their control systems, generating Uttal standards of behavior (1983). According to Fernández (2005), to analyze the organizational culture, four elements must be taken into account: the environment, the organization's value system, the historical form of its constitution and the functions that it fulfills for an organization from which it achieves its objectives and defined purposes.

In the same way, Schein (2004) defines the levels of the culture by affirming that the first one is composed of artifacts and creations (in this one culture is manifested but not the essence), the second one by the values and beliefs that guide the behavior of the members of the organization (in this, the staff has a greater knowledge of the culture) and the third is composed of the basic assumptions where people have the ability to perceive, think, feel and act (this is the deepest) (see Figure 1).

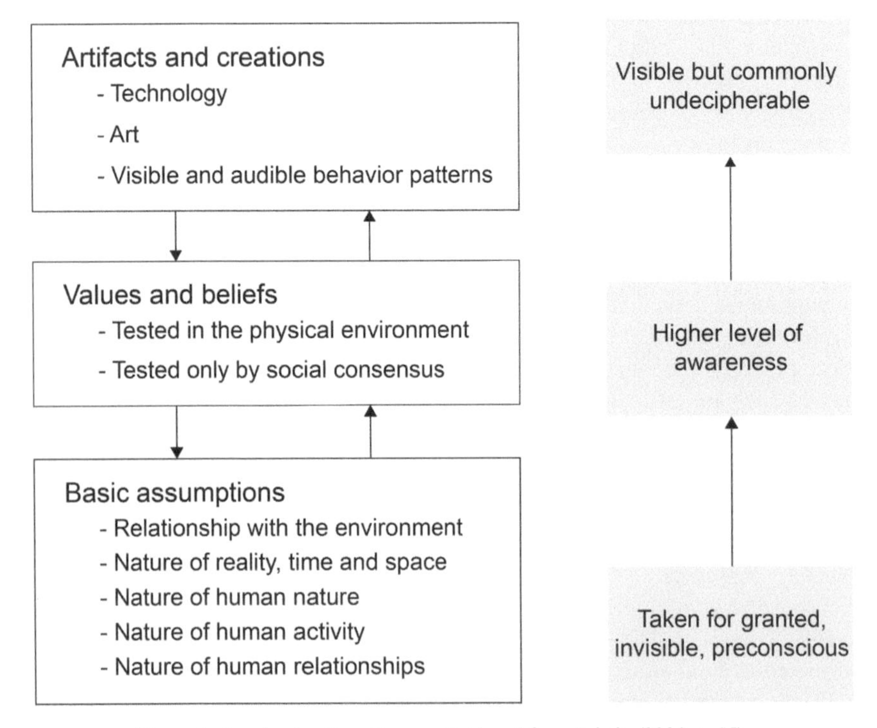

Figure 1. Levels of culture. Source: Retrieved from Schein (2004, p. 55).

Characteristics and Functions of the Organizational Culture

Regarding the characteristics and functions of the organizational culture, authors such as Luthans, Wheelen, Hunger, Sánchez and Robbins, among others, may have identified and explained them as a starting point for their definition, so these statements coincide in certain aspects or complement each other.

According to Luthans (2010), organizational culture has six characteristics:

1. Regularity of the observed behaviors: The members of an organization use the same terminology, language, rituals, which are related to the behavior.
2. Standards: Refer to regulations, behavior guidelines and work policies to do things in the organization.
3. Dominant values: Those that must be shared and defended by the members of the organization.
4. Philosophy: The policies that establish how employees and customers should be treated.
5. Rules: Established guidelines aimed at behavior within the organization. People who join the organization must learn and adapt to them.
6. Organizational climate: Feeling that is transmitted by the interaction of people, the treatment of customers and suppliers among others.

The characteristics have different measures and therefore it is feasible that controversies are generated. In each organization, these characteristics may have a different scale from the traditional and autocratic style to the participatory and democratic one.

On the other hand, Wheelen et al. (2007) set out that organizational culture is characterized by intensity and integration. Intensity is understood as the acceptance that the members of the organization have regarding standards, values or other components of culture so that when the organization has an intensive culture, the members of the organization have similar behavior. Regarding integration, it refers to the fact of sharing a culture that is common to all members of the organization. Finally, the dominant culture is presented in military institutions due to its pyramidal structure and power allocation.

Additionally, Robbins and Judge (2009) mention that organizational personality results from the way employees perceive the characteristics of an organization's culture, such as:

1. Innovation and acceptance of risk;
2. The attention to detail in the production;
3. Orientation to results rather than procedures;
4. The orientation to the people more than to the work that they develop;
5. Orientation to teamwork rather than individual work;
6. Aggressiveness, i.e., competitiveness versus comfort and stability;
7. Stability.

According to the aforementioned functions, the organizational culture influence people and that is why modifying strategic planning can become a complex situation. In that sense, the strategic direction of an institution will demand various changes and adjustments of the organizational constructs that make up its bases; in the case of the organizational culture, it will require, at the same time, its alignment and transformation, consistent with the previous changes and the development vision that is approved.

Robbins and Judge (2009) mention that the organizational culture has among its following functions:

1. Set limits, that is to say, that an organization of a particular form in addition to a demanding sense of belonging of the members.
2. Generate responsibility in the personnel of the organization oriented to the general and not particular good.
3. Contributes to stability.
4. Generate cohesion among the organization's staff.
5. Finally, lead employees according to their behavior and attitudes.

Types of Organizational Cultures

According to Cameron and Quinn (1999), the types of culture are classified as follows:

1. Clan culture: It is a very pleasant place to work in which people share personal information; as an extended family, the leaders or heads of the organization are seen as mentors and perhaps even as parent figures.
2. Adhocratic culture: It is a dynamic, business and creative place to work. The people in it strive and take risks. Leaders are considered innovators and risk-takers. Cohesion is based on the commitment to experimentation and innovation.
3. Market-oriented culture: It consists of a results-oriented organization, whose main concern is to do the job. The people there are competitive and goal-oriented. Leaders are demanding and competitive. The organization puts the emphasis on winning. According to Páramo (2001), a market-oriented organization is a foundation for the creation of an organizational culture in which marketing values predominate.
4. Hierarchical culture: It is a very formalized and structured place to work in which procedures govern what people do. Leaders take pride in being good coordinators and organizers of being efficient and intelligent. Compliance with rules and formal policies hold the organization together.

Bases of a Transformation of the Organizational Culture

During any process of cultural transformation, it is imperative that organizations understand that people constitute the main factor of change and that each of the dimensions that define the culture of an organization is closely connected with its collaborators, who are the main asset of any business.

On the other hand, the premises of a model that allows transforming the organizational culture, with the consequent benefits to the business management system, have to do with:

- A participatory leadership style in transit to transformational.
- Proactive results-based management system, supported by management indicators that allow relating the impact of cultural transformation on the creation, socialization, appropriation and use of advanced operational practices.
- The continuous improvement of all products, technologies, processes and practices.

In any institution, it is essential to create the bases of transformation on solid principles and values so that the progressive advances are achieved in addition to having the strongest support from the direction and have the support and commitment of all the collaborators of the organization.

Companies without exception, whether intentionally or not, create an organizational culture, which characterizes them and makes them stand out that can determine the most important decisions that are taken, the way things are done and the relationships between collaborators to fulfill the objectives and goals set. Corporate success or failure can be associated with some kind of predominant culture, which in turn is related to the level of participation and commitment of employees in the fulfillment of organizational purposes.

When employees appropriate the changes and have direct participation in the future of the company, they become the most valuable assets to promote cultural change. It is important to involve, communicate and facilitate the mechanisms that make participation effective. Receiving feedback on the progress of the process, being consulted and heard must be inviolable principles of the process.

Cultural Transformation in Alignment with Corporate Sustainability Management

The implementation of a management system focused on sustainability represents major challenges for transforming the organizational culture and together with this the design of an appropriate strategy to manage resistance to change. Organizations that seek to reach their full potential in the capacity and resources demanded by the sustainability management system must prepare to stand before major barriers, many of them determined by human and organizational constraints.

Within the organizational culture, the culture of sustainability can be one of its strongest or most responsible assets.

Working to transform a culture of sustainability into an organization with the purpose of aligning it with complexity and greater effectiveness of the management system will require profound changes in the way that have been working, the values and many of the ethical principles that guide the action.

Various theories have expounded the relationship between culture and sustainability and similarly sustainable development (Throsby 2017).

It may be more than a presumption or an argument that organizations that have worked intentionally to build their own culture that aligns with the strategic direction of their management systems, enjoy one of the competitive advantages that are difficult to imitate and may require a lot of time from other organizations to achieve the changes that have taken place.

An organization that works intentionally to develop a culture of sustainability is one in which employees share values and beliefs about the need to raise awareness related to economic, social and environmental responsibility, consciously fostering a healthy environment for employees, the community and other stakeholders, while dedicating sustained efforts to operate successfully in a sustainable manner.

Organizations that meet high standards of sustainability in addition to showing long-term work expectations meet different characteristics from organizations that have not adopted policies focused on the development and formative growth of human talent, leading to consolidate the changes that the transformation of processes and practices will impose to achieve new goals of sustainability with organized procedures for the active participation of stakeholders.

The main recipe for the leaders of the organization is to develop a strong, highly inclusive organizational culture oriented toward sustainability, which permeates and unites corporate members and fosters a sense of identity and commitment to common corporate environmental goals and aspirations (Dodge 1997).

The need to integrate the organizational culture and corporate sustainability can lay the following foundations and expectations in the collaborators:

1. Share similar attitudes, practices and values in the cultural approaches that are integrated and aligned to improve performance.
2. A stronger argument to improve efficiency and rational use of resources.
3. Understanding of phenomena and realities that go beyond the corporate framework or environment and integrate us more quickly to communities and society in general.

The collaborators whose work align with various types and developments of organizational culture will be interested in dedicating greater efforts to sustainability aspects that more easily fit the dominant culture and that can focus on the development of human talent, efficiency in the use of resources, greater alignment with customer requirements and expectations, participation and commitment in decision-making, including environmental protection.

By recognizing that there can be different types of culture, it is possible to understand that different subcultures coexist in companies by work areas, group or work team and each member of the various subcultures can express dissimilar attitudes toward corporate sustainability.

The challenge of each organization to be more sustainable has allowed the identification of a series of barriers and limitations against the purposes of working for sustainability, which is related to poor cultural change, lack of organizational flexibility, leadership style, commitment and participation and the inability of management systems that support the sustainability system. There are very concrete practical actions that are lead very well by the top management and middle managers of an organization and can provide a favorable context for changes in the values and beliefs of the employees.

Apparently, the ideal profile of corporate cultural sustainability should be produced with the most appropriate mix of practices, values and beliefs that promote innovation, good communication, labor policies and flexibility that stimulate teamwork, awareness of the value involved in meeting objectives and principles of sustainability.

An organization's true commitment to sustainability is expressed in an organization's steadfast decision to dedicate resources to create an atmosphere that motivates employees to work in alignment with best practices, models and principles in ways that counteract the level of current ignorance and apathy in trying to adopt the most appropriate behaviors and attitudes.

Sustainability Management System

A management system is a logical and phased process made up of related rules and principles in an orderly manner, based on continuous improvement, and that seeks to contribute to the management of general or specific processes of an organization, which includes policy, objectives, organization, application, evaluation and strategies focused on the production of sustainable goods and services demanded by customers.

The system must be led and implemented by the employer with the active participation of the workers and/or contractors, guaranteeing through the system, the application of measures and strategies focused on improving the methods and practices of workers to produce sustainable goods and services. Similarly, it should also be able to encourage companies to avoid and mitigate negative impacts on society and the environment and make positive contributions.

Corporate Social Responsibility (CSR)

Applied sustainability or sustainable development implies in addition to an active and voluntary contribution to the improvement of the social, economic and environmental environment of companies, recently carried out with the aim of improving their competitive situation. It is a set of social obligations that a group, an individual, an organization or an institution can assume regarding the environmental and social environment in which it carries out its activity. Like sustainability, it goes beyond compliance with laws and regulations, assuming its strict respect and compliance (Corredera and González 2011).

The commitment that a company acquires before its stakeholders and society in general is related to the voluntary integration by companies of social and environmental concerns in their business operations and relationships with their interlocutors. Among other things, it involves contributing to sustainable development, security, health and social well-being, respect and compliance both local and international regulations, consider the interests and expectations of stakeholders and maintain ethical and upright behavior as well as transparency in its management (Corredera and González 2011).

According to Lizcano and Moneva (2004), the basic principles that constitute the fundamental rules that govern the socially responsible behavior of companies are transparency, materiality, verifiability, broad vision, continuous improvement and the social nature of the organization.

The SDGs must be understood as an agenda of concerns, that is, a battery of topics on which the company must identify whether due to the activity it develops, it affects its aggravation or, on the other hand, can contribute to its solution. The SDGs represent an incentive for organizations to be more ambitious and demanding in developing responsible behavior.

The CSR will allow progress in the fulfillment of the SDGs taking advantage of existing structures, while these represent an opportunity for the CSR function to value its contribution to the business mission.

Mission, Vision and Values

The mission is a short text (usually one to two sentences) that defines the company's reason for being the fundamental purpose of the entity and especially what it provides to customers. The mission statement should inform executives and employees about the general objective they should pursue together (Kaplan and Norton 2008).

The vision must be very brief, simple and easily understood by all in order to be able to use it as a communicational message to motivate the employees of an organization and transmit to the market what they want to be (employees) (Kovacevic and Reynoso 2010)

The values of a company define its attitude, behavior and character (Kaplan and Norton 2008).

Stakeholder Management

Stakeholders are all those people or organizations that affect or are affected by the project, either positively or negatively.

According to the Project Management Institute (PMI), the stakeholders are those people, group or organizations that can affect, be affected by or perceive themselves to be affected by a decision, activity or result. Among those interested (stakeholders), the following can be mentioned (Project Management Institute 2017):

- Sponsor
- Customers
- Users
- Sellers
- Suppliers or contractors
- Business partners
- Organization groups
- Functional managers
- Other interested parties.

Stakeholder management involves processes of systematic identification, stakeholder analysis, design, planning and implementation of actions that allow managing the various requirements and expectations of interested parties.

Strategic Diagnosis, SWOT Analysis (An Acronym for Strengths, Weaknesses Opportunities and Threats)

In formulating the vision, the company has a clear picture of what it needs to achieve. Now it performs an external and internal analysis that includes a comprehensive evaluation of its own capabilities and performance in relation to its competitors as well as its positioning with respect to industry trends (Kaplan and Norton 2008)

External Analysis

The external analysis evaluates the macroeconomic environment of economic growth, interest rates, exchange movements, prices of production factors, regulations and general expectations of the role that the organization has in society (Porter 1980).

Internal Analysis

The internal analysis examines the performance and capabilities of an organization; it is the study of the key elements that condition certain strengths and weaknesses that are subject to its resources (Kaplan and Norton 2008).

SWOT Analysis

Once the external and internal analyzes have been completed, a SWOT analysis is developed, which is very likely to be the first and most important of all strategy analysis tools to identify the company's current strengths, weaknesses, emerging opportunities and worrisome (Kaplan and Norton 2008). With the help of the SWOT analysis, a company can analyze what it can do today, the organization's strengths and weaknesses and what it could do in relation to the external environment, the opportunities and external threats (Olve et al. 2002).

External and internal analyzes should be developed with sufficient scope to consider environmental, social and economic aspects and be guided on the basis of a systematic and exhaustive review of the activities, products and services of the organization that have (or may have) internal and external environmental, social and economic impacts.

The diagnosis should be promoted as a participatory process that should include all stakeholders of the organization.

Some of the aspects that should be known in this phase of the management process are to what extent the organization and its stakeholders are familiar with the SDGs, which will be evaluated based on the level of knowledge of the objectives and its goals, responsibilities that concern each area of work and worker as well as the threats and opportunities prevailing in the broad context of work.

As part of the diagnosis, priority SDGs, processes, technologies or products intended for compliance with the SDGs should be identified as well as which areas of their activity could be improved. Again, consultation with all stakeholders is necessary.

The Sustainability Policy

A policy is a guide for decision-making that links the formulation of the strategy with its application.

The sustainability policy provides a direction to the efforts of senior management as well as a frame of reference to establish the objectives and take appropriate actions to achieve the expected results.

The sustainability policy could be based on the following commitments:

- Compliance with environmental legislation and other requirements that the organization subscribes.
- Continuous improvement of the management and good environmental, social and economic practices of the organization.

Management System Objectives

The objectives are the expression of the achievements that the organization expects to reach within a certain period.

They must be consistent with the management system policy and strategy; its main feature is to measure itself in terms of efficiency, efficacy and effectiveness. The question to answer is:

- Where do we want to go?
- What results do we hope to achieve?

Once the diagnosis is finished, it is essential to establish the objectives in such a way that the SDGs can be integrated into the strategic business objectives.

At this point, the role of senior management, its commitment and support in the definition of appropriate indicators and goals are relevant, leaving for granted the importance of reference levels for each stage.

Organizational Structure

According to Miles et al. (1978), an organization is both its purpose and the mechanism built to achieve the purpose, that is, the organization encompasses both the objectives and all the elements that represent the combination of specific elements.

On the other hand, Miles et al. (1978) affirm that the design and processes that take place within the organization are closely aligned; it is difficult to talk about one without mentioning the other. Accordingly, these authors illustrate how design is interconnected with concepts such as leadership and communication, and how this mutual connection influences the processes of an organization's life cycle.

On the other hand, Bloisi et al. (2007) state that the organizational structure is defined as a grouping of people and tasks in different units to promote coordination of communication, decisions and behaviors. By considering the close connection between the processes that take place within an organization, it becomes easier to understand the complexity of the tasks and how to manage them efficiently.

Additionally, Daft (2015) refers that the organizational dimensions are divided into two types: structural and contextual. The structural ones provide labels to describe the internal characteristics of an organization, while the contextual ones characterize the entire organization, including its size, technology, environment and goals. The organization chart is the visual representation of a complete set of implicit activities and processes in an organization.

For Daft (2004), the organizational structure defines what power is like, the way responsibilities are assigned and how work tasks are divided, grouped and coordinated within an organization. In this way, the structure is first defined and a general description of the structural design is provided; then a perspective for sharing the information is established (how to design the vertical and horizontal links in order to provide the necessary information flow).

Decentralization or Centralization

Depending on the size of the organization, managers can use decentralized techniques to manage, direct and make decisions in the companies, which increases the difficulty in control and management. A decentralized organization is one where decision-making is not limited to a few senior managers but should include managers at different managerial considering important operational decisions (Chen et al. 2015).

A pending question in many modern organizations is to what extent the decision-making power must be centralized or decentralized.

Bloisi et al. (2007) state that centralization is the concentration of authority and decision-making from the top of the organization and decentralization is defined as the distribution of authority and decision-making units in an organization.

The commitments and work that will require compliance with the SDGs will force organizations to seek the best management and optimization strategies in resource management; in this sense, one of the first actions is to integrate sustainability into the fulfillment of its great goals, adopt the most flexible organizational structures with agile and decentralized decision-making as well as adopting timely management strategies throughout the value chain of the product or delivery service.

Lastly, it will be necessary to build the most effective information mechanisms with all interested parties so as to enable it allows timely decisions on development goals and common interest regarding the SDGs.

Leadership

Regarding leadership, Northouse (2013) mentions that this is a process by which an individual influences a group of individuals to achieve a common goal. Meanwhile, Bunmi (2007) refers that leadership is a process of social influence in which the leader seeks the voluntary participation of subordinates in an effort to achieve the objectives of the organization.

Additionally, today's organizations need effective leaders who understand the complexities of the rapidly changing global environment. If the task is highly structured and the leader has a good relationship with the collaborators, the effectiveness of the work will be high. In research, it has been known that leaders are very careful to involve all team members in any discussion and, on the contrary, tend to work with a small but highly motivated team.

For Voon et al. (2011), in the relationship between organizational culture and leadership styles, factors such as charismatic leadership, transformational leadership and transactional leadership are used.

According to Ismai et al. (2009), the transformational leadership style focuses on the development of followers as well as their needs. Therefore, managers with a transformational leadership style try to focus on the growth and development of the employees' value system, their inspiring level and morality as the preamble of their skills.

Meanwhile, Trottier et al. (2008) affirm that the transactional leadership style is based more on exchanges between the leader and the follower by which the followers are compensated, thanks to the fulfillment of the objectives or specific performance criteria. The transactional leader will first validate the relationship between performance and reward, then redeem for an appropriate response that encourages subordinates to consistently improve performance.

Charismatic leadership is, without a doubt, one of the most successful leadership styles. In this style, leaders develop a vision and supporters must follow and execute the vision; there, innovation and creativity are invited, and the leader considers himself motivating for the collaborators.

The biggest drawback of this leadership style is that the supporter is totally dependent on the leader and once the leader leaves the organization, they see themselves without direction. The problem worsens as charismatic leaders do not train their subordinates to act as their replacements in the future. This style of leadership results in happy followers but a few cases of future leaders. Therefore, it can have a long-term negative effect on organizational performance (Germano 2010).

Leadership and Commitment to the Management System

Senior management must demonstrate leadership and commitment in the implementation and proper functioning of management systems (International Organization for Standardization 2016) with greater visibility in systems, programs and projects of such a wide and complex scope as that related to sustainability.

- Take full responsibility and accountability for all attitudes, practices and related to sustainability.
- Establish a sustainability policy, aligned with the strategic objectives of the organization.
- Integrate the different requirements of the management system with the company's business processes.
- Direct the necessary resources to work toward the fulfillment of sustainability objectives and goals.
- Communicate in a consistent and sustained manner the importance of sustainability management efficiently and in accordance with the requirements.
- Ensure that the management system achieves the expected results.
- Lead or support the people who work based on the achievement of greater efficiency of the management system.
- Promote and encourage continuous improvement.
- Support the relevant management roles, and the leadership applied to their areas of responsibility must be demonstrated.
- Develop, lead and promote a culture of sustainability in the company that supports the objectives and expected results.
- Stimulate the report of situations that represent threats and opportunities to achieve sustainability goals and objectives.
- Establish consultation processes and employee participation.
- Create sustainability committees according to priority SDGs and ensure their full operation.

Leadership, Ethics and Social Responsibility

- Leaders can contribute significantly to the success of management systems and the standardization of good work practices in a company. Even more so if these practices are firmly connected to superior performances and of high strategic significance from the competitive point of view.
- In strategic management, the leader plays different roles: introduces the environment for change, creates the leadership team by selecting key players in the organization and formulates the vision and strategy with the help of a visionary process that clarifies the strategy for the understanding of the entire organization (Moesia 2004)
- The leader in an organization not only values a context, plans the vision and objectives but also designs the strategies and manages the operational activities, organizing the collaborators for their effective execution and adapting it to the demands of the organization's development.
- The leader is responsible for transmitting energy, generating commitments, increasing the morale and learning of the work team in a context of fruitful relationships with all stakeholders in a permanent analysis of legitimacy, urgency and power.
- Regardless of the leadership style that prevails in an organization or certain areas, it cannot act outside the ethics and principles that define management and good practices.
- According to Álvarez et al. (2001), the leader must distinguish himself from the rest of the employees of the organization for his commitment to use in his daily work; the qualities inherent to the human being are respect, generosity, patience, dignity, the ability to listen and humility.
- Ethical leadership must include vision, coordination and change, based on the moral virtues of prudence, justice, strength and temperance.

The Value of Supervision and the Supervisor

The word supervision is derived from two Latin words 'super', which means above, and 'videre', which means see. A supervisor is someone who sees or observes from above a group of people, their team of employees or subordinates and helps them to develop a job designed to satisfy with quality the needs of a client.

The supervisor is the key man in a sustainability program because he is in permanent contact with plant or area employees.

Being a supervisor means being in full view of everyone, is somebody who has a better position to see others and is also the most visible. He is a notorious figure, and he is surrounded by spectators, people who observe him, expect something from him and depend in some way on what he does or does not do (Jauregui 2006).

The leadership style and behavior in which a person who works in a key position is important to build a culture of adequate sustainability. In this sense, leadership is the ability to foster good relationships with workers. Leadership theories as a leader-member exchange argue that good follower-leader relationships encourage followers to behave in a way that is aligned with the leader's goals and values (Hofman and Morgeson 1999).

If the director has a good relationship with the staff and they behave in a way that promotes best practices of sustainability at work, employees will feel inclined to adopt the most appropriate work practices as a form of reciprocity. In this regard, it is important that managers try to develop a leadership style that promotes collaboration and relationship building while maintaining authority and discipline.

Transformational leadership is presented as a new approach to leadership-oriented toward participation and flexibility in the organization (Bryman 1996). This type of leadership requires the involvement and power of an expert to cause changes in followers about their beliefs and values.

One of the biggest challenges facing any frontline supervisor is to motivate the employee to use advanced work practices that promote activities specifically focused on achieving new and superior sustainability goals, like those listed below:

1. The supervisor can develop positive attitudes toward sustainability with a personal example.
2. The supervisor must develop the bases to obtain the greatest cooperation of the workers through good personal relationships.
3. The supervisor must properly treat the employee who does not show the best attitudes to contribute to the collaborative work that the purpose of sustainability demands, relying on formal education processes and/or the execution of activities.

Supervisory duties involve technical and behavioral knowledge. The supervisor stands as a fundamental factor in achieving the company's sustainability goals. His management leads the process by offering the worker the necessary advice to develop the habits that support the management system and the best work practices.

Supervisory functions according to sustainability purposes and goals:

1. Assess the capacities/competences of people according to the assigned tasks.
2. Analyze the relationship between the higher goals set for sustainability management and the capacities and resources assigned to achieve them.
3. Implement effective controls and/or precautions to prevent negative deviations, such as inspections, detection controls, information, training, procedures and supervision.
4. Improve and simplify permanent labor practices that aim to improve sustainability levels.
5. Inform and instruct on good practices.
6. Demonstrate the correct procedures.
7. Assign the tasks according to the person's capacity and perform direct supervision until they reach the level of competence.
8. Verify those correct procedures are understood, and precautions are taken to preserve achievement and reach new goals.

9. Promote scenarios and workspaces to reinforce learning.
10. Listen, consult and share experiences with employees.
11. Inform new employees about ways to get help and advice in their absence and what they should do if they have doubts.
12. Provide any written procedure for the best execution of the task.
13. Instilling values, explain the importance of sustainability, the organization's commitment to promote and encourage the right attitudes, raise awareness and set a good example
14. Motivate properly, facilitating processes and guiding.
15. Control, organize and discipline if necessary.

Strategies, Indicators and Goals

Strategies are the guidelines that help to choose the appropriate actions to achieve the goals of the organization, accompanied by indicators, goals, programs and action plans, which finally strategies the basis for establishing priorities in the allocation of resources.

Furthermore, the strategy is how a company strives to create value, develop a unique advantage and capture maximum market share if it is in the company's business interests.

The strategies are related to the actions that a company plans or implements to achieve the long-term objectives. The union of all these actions forms the strategic plan of a company. The strategic plans emanating from these involve the participation of all levels of management and collaborators of the company. The strategies try to answer the following question:

How can I achieve the objectives or how can I reach the expected results?

When organizations through diagnosis have decided their priorities in relation to the SDGs have built their development and integration objectives with the criteria of all interested parties, they should work focused on designing and implementing the best strategies in the context of extensive programs, projects and work plans that will allow specifying the necessary activities and actions.

According to the SDGs, companies may think of strategies that place their emphasis on the following scopes:

SDG 1: No Poverty

New employment opportunities, improvement of training levels and improvement of employment with the consequent wage increase, a decrease in the flexible wages and an increase in salary in kind with clothing, household items, food, bonuses or even with intangible goods and services such as accommodation or home classes.

SDG 2: Good Health and Wellbeing

In this direction, organizations have an enormous responsibility with the promotion of healthy habits, the prevention of accidents and occupational diseases, the provision and allocation of resources that allow managing the most appropriate strategies related to the design, implementation, review, evaluation and improvement of strategies to prevent and control.

SDG 3: Quality Education

Companies can be involved in this SDG through employee training and education in consumer and citizen habits and values. Likewise, there are many other issues that create a strong link between quality education, welfare or end of poverty, and have to do with:

- Promote quality education through social action programs aimed at groups at risk of social exclusion, such as immigrants, women who have suffered gender violence or inclusive education for people with disabilities.
- Programs that contribute to the professional enrichment of employees aimed at improving job skills, project management, innovation, entrepreneurship, lean six sigma, etc.
- Digital education and digital connectivity improving people's quality of life and facilitating access to knowledge.
- Corporate university to align with the growing demands of the organization and develop new processes and tasks with higher employability expectations of the workforce.

SDG 4: Gender Equality

Regarding equality, the company must work to reduce the wage gap and to promote the promotion of female human capital in order to increase motivation and the possibilities of evolving in the business environment. Empower them with greater capacity to make decisions aligned with greater skills training and job skills training.

SDG 5: Clean Water and Sanitation

Companies must introduce strategies to avoid depletion and contamination. The calculation of the direct water footprint in an organization has become an instrument that allows knowing the impact of business activities on the use and exploitation of water resources.

Companies that, due to their processes, for instance, the one that belongs to the construction sector by the emission of particles, others by fats or oils, but there are an important group of process industries that do so from their packaging for their products.

Not less important is the work of prevention, detection and maintenance to reduce leaks in the supply network, using early alarm systems, periodic reviews, attention to contingencies, etc., and promoting innovation in treatment systems for wastewater reuse.

SDG 6: Affordable and Clean Energy

Companies will have to invest in renewable energies and technologies that improve energy efficiency, reducing consumption and pollution.

This should be supported in the wider and more widespread use of natural lighting, the implementation of more efficient lighting systems and sources, the use of presence sensors in the sites to be illuminated, increased awareness of rational use among employees and appropriation of new cleaner sources of power generation.

SDG 7: Decent Work and Economic Growth

This SDG requires the company to create jobs, eradicate forced and child labor and create safe and healthy work environments. The transformation of the hiring models to more guarantors in health and wellness.

Enriched work as part of positions with wide profiles, where autonomy, flexibility predominates, and innovation is promoted. Awaken the entrepreneurship that finally contributes to the economic improvement of the company and society.

SDG 8: Industry, Innovation and Infrastructure

Infrastructures, industrialization and innovation allow companies to develop their activity with higher levels of efficiency and effectiveness.

Stimulating research initiatives as part of innovation processes to promote sustainability must be the sustained responsibility of organizations that through explicit policies and strategies, channel and receive all institutional priority.

Organizations should stimulate close relationships with universities, and in this way improve everything they do, renewing the old, replacing inefficient technologies or that significantly affect sustainability.

SDG 9: Reduced Inequalities

Equality policies, the labor integration of people at risk of social exclusion, the reduction of wage inequalities, the commercialization of affordable products and services and good tax practices are some of the tools of companies to reduce inequality. The inclusion of technologies and the appropriation of excellent practices in the best management of people with disabilities, with gender or generational differences, etc.

SDG 10: Sustainable Cities and Communities

Sustainable corporate buildings have a great environmental impact and directly on essential aspects, such as energy consumption and the use of means of transport. The development of mobility plans for employees, offering concrete measures and promoting sustainable transport.

SDG 11: Responsible Consumption and Production

Business models will have to evolve to coincide with the concept of a circular economy, which minimizes resource consumption and waste generation. Inclusion and innovation of more efficient technology. Minimize emissions in product purchase and marketing and circularity in their processes to preserve resources.

SDG 12: Climate Action

The challenge of companies to set ambitious targets, accurately measure their emissions throughout the entire value chain and incorporate climate change adaptation and mitigation initiatives alongside new technological developments can help companies reduce greenhouse gases.

SDG 13: Life Below Water

Reduce the production and consumption of plastic as an input, raw materials, packaging and unit load device. Encourage the use of waste as a resource, reuse and recycle as much as possible.

SDG 14: Life on Land

Natural habitat management is essential to guarantee the continuity of resources, maintain the natural balance of ecosystems and counteract the effects of climate change.

Acquisitions and purchases of raw materials or office supplies from suppliers that classify as sustainable sources that demonstrate through their practices, culture and image that production is responsible for the environment. Develop good recycling practices.

SDG 15: Peace, Justice and Strong Institutions

Transparency, good governance and regulatory compliance practices, respect for human rights, due diligence processes or mechanisms for dialogue and complaint will allow the private sector to combat injustices and corruption. Good management practices of stakeholders or interest groups can become an effective instrument to implement the above strategies, including the various dialogue mechanisms and an effective way to block corruption actions. Implementation of ethical codes.

SDG 16: Partnerships to Achieve the Objectives

The creation of alliances includes both the mobilization of economic resources and the exchange of knowledge, technical capacity, technology and human resources.

Public-private alliances with local communities, universities, companies of the same sector, etc., in search of solutions to complex problems and encourage research. Participation in conferences and events that promote the exchange.

Design and Implementation of Key Performance Indicators (KPI)

The design and implementation of key performance indicators is a stage related to the definition of objectives as the basis for guiding, monitoring and communicating progress.

Some companies establish general or vague goals that do not allow themselves to measure progress. In these cases, the recommendation is to select several KPIs and that each one conforms to the basis for a specific, measurable and time-limited objective.

The ideal starting point for the selection of KPIs is the range of indicators used to assess the impacts, each company can reduce the selection to some key indicators that best express its impact on the issue of sustainable development. Some of the most important indicators from the environmental, economic and social scope are proposed below.

Environmental

- Natural gas, electricity and water consumption.
- Emissions.
- Used water.
- Biological and chemical oxygen demand.
- Water temperature.
- Waste recovered or disposed of.

Economic

- Rentability.
- Value delivered to customers.
- Financial strength.
- Investor relations.
- Risk and crisis management.

Social

- Work accident rate.
- Occupational disease rate.
- Number of employee training programs.
- Number of non-profit programs.
- Unsafe working conditions controlled.
- Unsafe work behaviors controlled.
- Medical disability days were reduced.

Conclusions

Business practice evidence that leadership and culture in sustainability management are two important predictors of good performance of management systems, considering that leadership in sustainability focuses on the process of monitoring compliance with the objectives system management process, adjusting any significant deviation through the authority of supervisors to enforce the rules of its application, the control of commitment and consistent deployment of good sustainability practices.

Among all the elements that contribute to better management of the indicators that measure the performance of the sustainability management system, the commitment with the application of strategies and good practices of high-level managers, middle management and supervisors are the most important.

The adoption of culture models that are oriented to strengthen the changes and transformation processes of sustainability will require a guiding axis throughout the process, where consistent well-planned actions and broad integration of the efforts deployed by each organization prevail, with the intention to perceive of all opportunities for improvement and options to promote innovation processes and individual and collective responsibility for managing sustainability.

The pillars are proposed in the chapter for organizations to work in an aligned and organized way in the processes of evaluation, management and improvement of sustainability, reinforcing the procedures, technologies, practices or behaviors as the basis of the management process, connecting changes with organizational culture features through personal and practiced values, myths, rites, beliefs, etc., and promoting the most appropriate management styles to achieve the objectives.

References

Altieri, A. 2001. ¿Qué es la cultura? La lámpara de Diógenes. Revista semestral de filosofía, 2(4), 15–20. Retrieved from http://redalycuaemex.mx.

Álvarez, L., Santos, M. and Vázquez, R. 2001. El concepto de orientación al mercado: Perspectivas, modelos y dimensiones de análisis. Departamento de Administración de Empresas y Contabilidad, Universidad de Oviedo, España, 245, 1–28. Retrieved from https://dialnet.unirioja.es/servlet/articulo?codigo=1252647.

Belalcázar, S. B. 2012. Cultura organizacional. Informes psicológicos, 12(1), 41–51. Retrieved from de https://revistas.upb.edu.co/index.php/informespsicologicos/article/view/1815/1747.

Block, L. 2003. The leadership-culture connection: an exploratory investigation. Leadership and Organization Development Journal, 24, 318–34.

Bloisi, W., Cook, C. W. and Hunsaker, P. L. 2007. Management and Organizational Behaviour. 2da. edition. Maidenhead: McGraw-Hill.

Bryman, A. 1996. Leadership in organizations. *In*: Clegg, S. R., Hardy, C. and Nord, W. R. (eds.). Handbook of Organization Studies, Sage, London.

Bunmi, O. 2007. Effect of leadership style on job-related tension and psychological sense of community in work organizations: a case study of four organizations in Lagos State. e-Journal of Sociology, 4(2), Nigeria, Bangladesh.

Calderón, G., Naranjo, J. and Álvarez, C. 2010. Gestión humana en Colombia: Sus características, retos y aportes. Una aproximación a un sistema integral. Cuadernos de administración, 23(41), 13–36.

Cameron, K. and Quinn, R. 1999. Diagnosing and Changing Organizational Culture: Based on the Competing Values Framework. New York. Addison Wesley Publishing Company.

Corredera, J. and González, M. 2011. Diccionario LID Responsabilidad y sostenibilidad. Madrid: LID.

Chen, C., Pan, F. and Wang, Y. 2015. Determinants and consequences of transfer pricing autonomy: An empirical investigation. Journal of Management Accounting Research, 27(2), 225–259.

Daft, R. L. 2004. Organizational Theory and Design. Cincinnati, OH: Southwestern.

Daft, R. L. 2015. Teoría y diseño organizacional. México: Cengage Learning Editores.

Dodge, J. 1997. Reassessing culture and strategy: Environmental improvement, structure, leadership and control. pp. 104–126. *In*: Welford, R. (ed.). Corporate Environmental Management. 2. Culture and Organizations. London: Earthscan.

Erdogan, B., Liden, R. C. and Kraimer, M. 2006. Justice and leader-member exchange: The moderating role of organizational culture. Academy of Management Journal, 49, 395–406.

Fernández, E. 2005. Estrategia de innovación. Madrid, España: Thomson editores.

Flamholtz, E. G. and Randle, Y. 2012. Corporate culture, business models, competitive advantage, strategic assets and the bottom line: Theoretical and measurement issues. Journal of Human Resource Costing and Accounting, 16(2), 76–94.

Germano, M. A. 2010. Leadership Style and Organizational Impact. Retrieved from http://alaapa.org/newsletter/2010/06/08/spotlight/.

Global Sustainable Development Report. 2019. The future is now science for achieving sustainable development. https://sustainabledevelopment.un.org/content/documents/24797GSDR_report_2019.pdf.

Hall, E. T. 1959. The Silent Language. Bantam Doubleday Dell Publishing Group, Inc. 1540 Broadway, New York.

Hofman, D. and Morgeson, F. 1999. Safety-related behaviours as a social exchange: the role of perceived organisational support and leader-member exchange. Journal of Applied Psychology, 84(2), 286–296.
International Organization for Standardization. 2016. Occupational health and safety management systems—Requirements with guidance for use (ISO/DIS Standard No. 45001. Retrieved from http://www.iso.org/iso/catalogue_detail?csnumber=63787).
Ismai, A., Halim, F., Munna, D., Abdullah, A., Shminan, A. and Muda, A. 2009. The mediating effect of empowerment in the relationship between transformational leadership and service quality. International Journal of Business Management, 4(4), 3–12.
Jauregui, A. 2006. Papel del supervisor. México: Editorial Pax.
Kaplan, R. and Norton, D. 2008. The execution premium: integrando la estrategia y las operaciones para lograr ventajas competitivas. España: Ediciones Deusto.
Kovacevic, A. and Reynoso, A. 2010. El diamante de la excelencia organizacional. Santiago de Chile: Aguilar Chilena de Ediciones S.A.
Kroeber, L. and Kluckhohn, C. 1952. Culture. New York: Meridian Books.
Kroeber, L. and Parsons, T. 1958. The concepts of culture and of social systems. American Sociological Review, 23, 582–83.
Loaiza, E., Salazar, P., Espinoza, L. and Lozano, M. 2019. Clima organizacional en la administración de empresas: Un enfoque de género. Revista científica mundo de la investigación y el conocimiento, 3(1), 3–25.
Lindell, M. and Karagozoglu, N. 2001. Corporate environmental behavior—a comparison between Nordic and US firms. Business Strategy and The Environment, 10, 38–51.
Lizcano, J. and Moneva, J. 2004. Marco Conceptual de la Responsabilidad Social Corporativa. Madrid: Asociación Española de Contabilidad y Administración de Empresas (AECA).
Luthans, F. 2010. Organizational Behavior. New York, Estados Unidos: Mcgraw Hill.
Miles, R., Snow, C., Meyer, A. and Coleman, H. 1978. Organizational strategy, structure, and process. Academy of Management Review, 3(3), 546–562.
Moesia, M. and Veldsman, T. 2004. The importance of different leadership roles in the strategic management process. S.A. Journal of HRM, 2(1), 26–36.
Northouse, P. G. 2013. Leadership: Theory and Practice. 6th edition. Thousand Oaks: SAGE Publications, Inc.
Olve, N. -G., Roy, J. and Wetter, M. 2002. Implantando y Gestionando El Cuadro de Mando Integral. Barcelona: Ediciones Gestión 2000.
Páramo, D. 2001. Hacia la construcción de un modelo de cultura organizacional orientada al mercado. Revista Colombiana de Marketing, 2(2), 1–26.
Porter, M. 1980. Competitive strategy: techniques for analizying industries and competitors. The Free Press.
Project Management Institute. 2017. A Guide to the Project Management Body of Knowledge (PMBOK Guide, 6e). Newtown Square, Pennsylvania: Project Management Institute, Inc. DOI: http://dx.doi.org/10.31095/podium.2018.34.6.
Real Academia Española. 2019. Cultura. Retrieved from https://dle.rae.es/?id=BetrEjX.
Robbins, S. and Judge, T. 2009. Comportamiento organizacional. Retrieved from https://psiqueunah.files.wordpress.com/2014/09/comportamiento-organizacional-13a-ed-nodrm.pdf.
Sánchez, J., Tejero, B., Yurrebaso, A. and Lanero, A. 2006. Cultura organizacional: Desentrañando vericuetos. Revista de antropología Iberoamericana, 1(3), 380–403. Retrieved from https://www.redalyc.org/articulo.oa?id=62310304.
Schein, E. 2004. Organizational culture and leadership (Vol. Third Edition). San Francisco: Jossy-Bass. Retrieved from http://www.untag- smd.ac.id/files/Perpustakaan_Digital_2/

ORGANIZATIONAL%20CULTURE%20Organizational%20Culture%20and%20 Leadership,%203rd%20Edition.pdf.

SDSN, Action Agenda Report. 2014. Public Consultation on Indicators for Sustainable Development Goals. https://www.unsdsn.org/news/2014/03/12/public-consultation-on-indicators-for-sustainable-development/.

Throsby, D. 2017. Culturally sustainable development: Theoretical concept or practical policy instrument? International Journal of Cultural Policy, 23, 133–147.

Trottier, T., Van Wart, M. and Wang, X. 2008. Examining the nature and significance of leadership in government organizations. Public Administration Review, 319–333.

Tylor, E. 1891. Primitive culture. Retrieved from: https://archive.org/stream/primitiveculture1891tylo#page/n3/mode/2up.

Uttal, B. 1983. The corporate culture vultures. Fortune, 66.

Voon, M., Lo, M., Ngui, K. and Ayob, N. 2011. The influence of leadership styles on employees job satisfaction in public sector organizations in Malaysia. International Journal of Business, Management and Social Sciences, 2(1), 24–32.

Wheelen, T., Hunger, J., Sánchez, M., Linde, G. and Mejía, H. 2007. Administración estratégica y política de negocios. Retrieved from https://aed1035gestionestrategica20181.files.wordpress.com/2018/02/administracic3b3n-estratc3a9gica-y-polc3adtica-de-negocios-wheelen-hunger-10ed.pdf.

Wehrmeyer, W., France, C. and Leach, M. 2010. Sustainable development management systems in global business organizations. Management Research Review, 33(11).

Chapter 7

Strategies to Improve Environmental Performance

Edilberto Miguel Avalos Ortecho

Introduction

For many years, companies around the world have been looking for different ways to be more competitive or to achieve competitiveness. Companies used to practice differentiation or implement low-cost strategies in order to be more competitive and profitable, but these measures are no longer enough. Instead, competitiveness now depends on the management of process production in order to minimize pollution, such as atmospheric emissions, effluents, and solid wastes. An additional challenge for companies is to deliver their customers' products that have better environmental performance and better value proposal to the market. Organizations must also factor external environmental impacts into their internal operations budget with the objective of being more competitive. On the other hand, companies that control operational pollution by preventive measures will gain social license (Zhang et al. 2015), and those that contribute to corporate social responsibility will qualify as sustainable. For these reasons, it is urgent that a strategy that makes companies more competitive and profitable by improving their environmental performance is found.

There are three ways to control environmental impacts: preventive measures (cleaner production), end of pipe technologies and a mix of both technologies. End-of-pipe technologies can help an organization comply with environmental regulations and pollution taxes, but it cannot provide an internal rate of return. On the other hand, companies that have altered their process performance to appeal to green markets and customers can achieve superior profits and better environmental performance by using resources efficiently in order to reduce pollution in a preventive way. As a result, it is possible to reduce operational costs and optimize profits. Therefore, ecologic consciousness appears to be a promising business strategy that can increase a company's bottom line. The question then follows is if there is a relationship

Operations and Process Department, Architecture and Engineering Faculty, Universidad de Lima, Av. Javier Prado Este N° 4600, Monterrico, Lima, Perú.
Email: eavalos@ulima.edu.pe

between industrial ecology as a strategy and the environmental performance of companies.

Industrial ecology as a strategy refers to taking action to prevent pollution, excluding corrective actions. Porter (1995), Hart and Ahuja (1996), Russo and Fouts (1997), Montabon et al. (2007) and Lopez-Gamero et al. (2009) agreed that there is a strong relationship between pollution prevention and environmental performance.

The primary objective of this chapter is to determine if there is a strong relationship between industrial ecology as a strategy (pollution prevention) and the environmental performance of companies. Another objective is to determine how to use environmental tools to reduce environmental impacts from process sources, i.e., in a preventive manner (by controlling raw materials and processes with clean technologies) and then use end-of-pipe technologies to address residual impacts, such as atmospheric emissions, effluents and solid wastes.

First, this chapter defines why companies must consider industrial ecology as a viable business strategy. Then, this chapter shows why environmentally conscious principles must be complementary with the study of tools, such as the environmental management system ISO 14001:2015, life cycle assessment system, e.g., ISO 14040, and monitoring environmental performance using indicators like ISO 14031 and the Sankey diagram with the objective of focussing on critical environmental problems. Additionally, this chapter discusses if there is a strong relationship between cleaner production and business performance of organizations using structural equation modeling with Amos software 22.0 and ChemCad 7.2. These models simulate processes in order to optimize environment process performance.

Finally, this chapter evaluates the hypothesis that the adoption of industrial ecology (pollution prevention) as a business strategy improves the environmental performance of companies.

Strategy

According to Porter (1996), "The myriad activities that go into creating, producing, selling, and delivering a product or service are the basis units of competitive advantage. Operational effectiveness means performing these activities better that is, faster or with fewer inputs and defects than rivals do. Companies can rear enormous advantages from operational effectiveness. The problem with operational effectiveness is that best practices are easily emulated." In this case, the competitive advantage exists only over a short term; for instance, a company implements environmental management system ISO 14001:2015 and is able to improve their environmental performance because they can manage a significant environmental indicator in a preventive way, and as a result, the company can reduce atmospheric emissions, effluents or solid wastes.

On the other side, by practicing continuous improvement, a company can reduce operational costs (direct costs, indirect costs and external costs) and consequently deliver greater value to customers or can create comparable value at a lower cost or do both. The trouble then arises on how a company could achieve a sustainable competitive advantage while preserving what is distinctive about that company. In other words, performing similar activities in one industry in different ways, using a different set of activities to deliver a unique mix of values. One of these different ways would be to control the quality of raw materials, for instance not using toxic materials because they do not create value for the product and they pollute the environment and increase production costs. Traditionally, organizations practiced generic strategies, such as cost leadership, differentiation and focus, but currently, these measures are not enough because they do not reduce pollution or wastes from source processes nor do they design products with a better environmental performance by delivering a sustainable product to the customer. In this way, it becomes clear that companies could adopt industrial ecology as a strategy to obtain superior sustainable performance among its competitors and achieve an optimized bottom line.

Industrial Ecology as a Strategy

"Industrial ecology (i.e.) is an emerging field, which covers the study of physical, chemical, and biological interactions and interrelationship both within and between industrial and ecological systems. Implementing an i.e. framework incorporates the strategy of cleaner production and pollution prevention in the industrial activities" (Basu and Van Zyl 2006). The aim of i.e. is to interpret and adapt to an understanding of the natural system and apply it to design of the manmade system in order to achieve a pattern of industrialization that is not only more efficient but that is intrinsically adjusted to the tolerances and characteristics of the natural system. The emphasis is on forms of technology that work with natural systems and not against them. Applied i.e. is an integrated management and technical program including:

- The creation of an industrial ecosystem.
- Balancing industrial input and output to natural ecosystem capacity.
- Dematerialization of industrial output.
- Improving the metabolic pathways of the industrial process and materials used.
- Policy alignment with a long-term perspective of industrial ecosystem evolution" (Tibbs 1992).

IE can be applied to any manufacturing or transformation sector like mining, textile, ceramic, cement and energy in order to seek out most opportunities for environmental improvement. IE as a strategy could be another way to improve process productivity (the creation of a unit and valuable position), and it reduces not only pollution but also operational costs. As a result, this will increase the value of the product for the customers and the organization (Avalos 2018). "Strategy is creating fit among a company's activities. The success of a strategy depends on doing many things well-not just a few-and integrating among them. If there is no fit among activities, there is no distinctive strategy and little sustainability" (Porter 1996). Accordingly, if companies adopt IE as a strategy, they should apply it not only to their operations but also to all other areas, such as marketing, finance, human resources, chain management, maintenance, quality and environmental systems and strategic planning by implementing a culture of pollution prevention. There must be operational, tactical and strategic actions that establish short-, intermediate- and

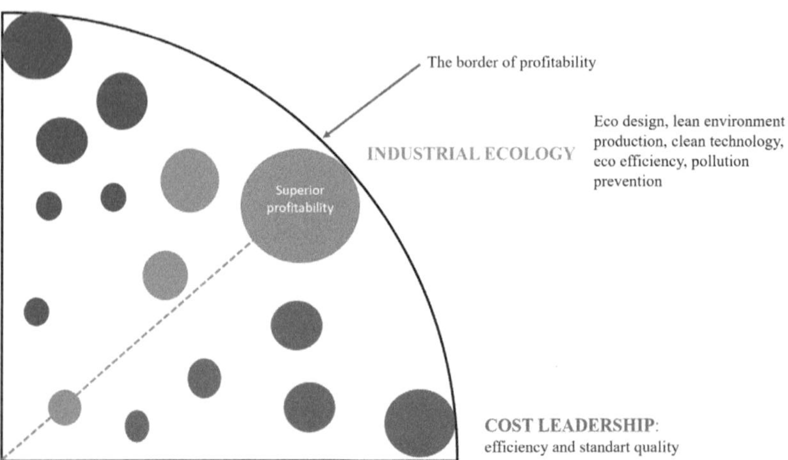

Figure 1. Comparison of strategies, with the border of the profitability shown (Avalos 2018).

long-term plans. As shown in Figure 1, IE as a strategy provides strong tools to companies, enabling them to increase environmental improvement by focussing on the source(s) of pollution, i.e., process production and quality of raw materials.

Industrial Ecosystem Model

In order to reduce or limit the environmental impact caused by human activities, it is necessary to increase the efficiency with which resources are consumed in the technosphere (Graedel and Allenby 2003).

"The model shows in Figure 2 how material as well energy is flowing between components in the technosphere, thus the energy supplier or energy generator has been added to the original industrial ecology model" (Despeisse et al. 2013). The proposal is to find a closed circuit in order to limit waste production.

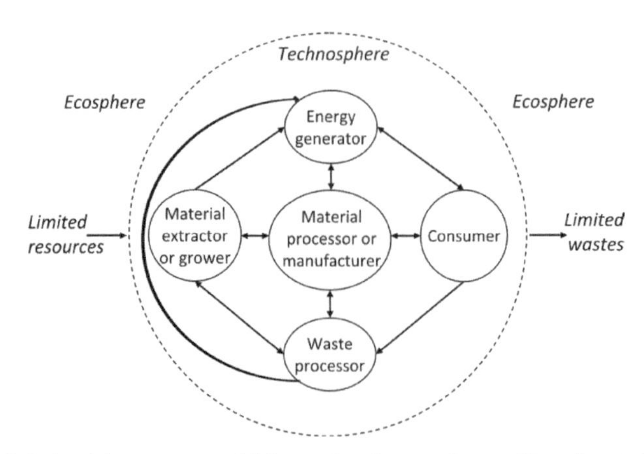

Figure 2. Industrial ecosystem model focussed on the manufacturer (Despeisse et al. 2013).

Manufacturing Ecosystems Model: From Linear to Quasi-Cycle Resource Flow Through a Process or System

Figure 3 shows process production as a manufacturing ecosystem where the main idea is to plan a closed production system in order to minimize waste. In this scenario, it is necessary to practice an operational strategy that helps the company manage pollution in a preventive manner (Avalos 2018).

"The data analysis should focus on the resource flows linking the systems' components and on how technologies are consuming and transforming the flows to create process maps of the manufacturing systems. As in an ecosystem, the importance is the overall productivity of resources and how they circulate in the system rather than the efficiency of individual processes or technologies" (Despeisse et al. 2013).

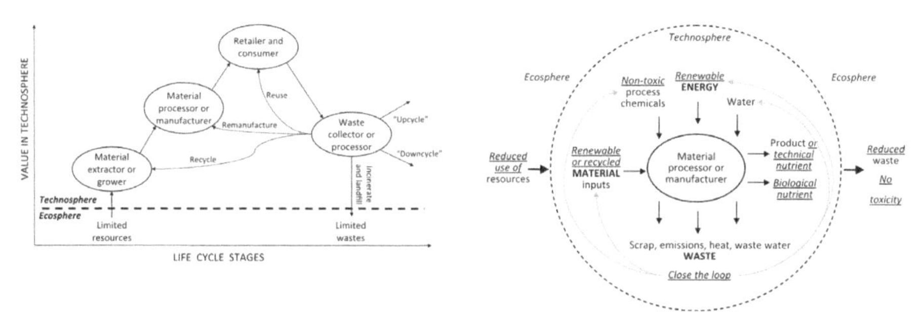

Figure 3. Manufacturing ecosystems model: from linear to quasi-cyclic resource flow through a process or system.

Manufacturing Ecosystems Model with Three Components and Material, Energy, Waste (MEW) Flows

As it shows on Figure 4 about manufacturing ecosystems model, in order to be more eco-efficient (energy, waste and material management), it is necessary to manage the resources aligned with lean environmental thinking, clean technology and an ecology strategy. The environmental aspect is managed with an environmental management system like ISO 14001:2015, which led companies to prevent pollution and standardizing production methods (Avalos 2018).

"By capturing the MEW (material, energy, waste) process flows systemically, potential interactions between processes can be identified to recover material and energy losses, 'capture' them and use them in another process. Using a systems view is a key element to move towards solutions, which bring opportunities to improve the system as a whole and avoid local, suboptimal solutions. This knowledge of the potential flow interactions, design methodologies can be developed to enable more environmentally sustainable manufacturing system creation" (Heilala et al. 2008).

"By taking a systems view, we acknowledge that everything affects everything else and therefore we need a tool which helps anticipating those effects while allowing the identification of long-term consequences and root causes. Using a process modelling approach to map MEW flows and analyse them can highlight

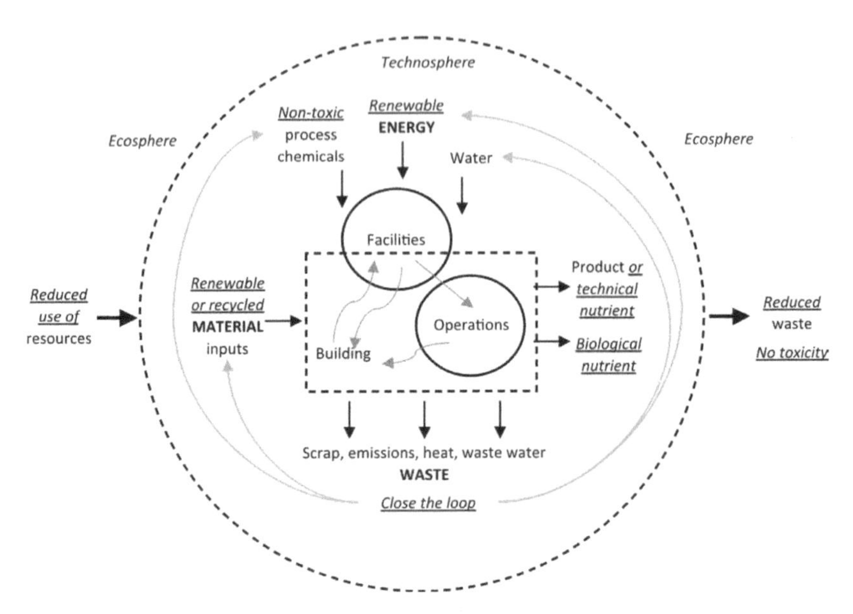

Figure 4. Manufacturing ecosystems model with three components and material, energy, waste (MEW) flows.

opportunities to use outputs from some activities to use outputs from some activities as inputs to others to reduce net consumption. Such principles for recovery and reuse of waste and energy, and therefore reduction of cost and environmental impact could be applied across a whole facility" (Heilala et al. 2008).

Manufacturing Ecosystem Example

This manufacturing ecosystem example suggests designing a manufacturing ecosystem in order to minimize not only the transformation process and operator movement but also the circulation time with the objective of reducing production cost and wastes (Avalos 2018).

Figure 5 represents a factory building and offices, the operation systems (composed of various processes) and the facilities supporting those processes and the building system. The type of improvement can be in organizational/operational management or technical/physical change, and the corresponding elements that should be targeted are resource flow and/or technology components. The generic examples below cover various MEW flows or technological solutions to identify improvements using simple rules based on recurring themes for industrial sustainability (adapted from Lovins et al. 2008).

Industrial Ecology: The Heart of Strategic Planning

Companies should consider IE as a strategy in their strategic planning in order to drive sustainable development. Economic and social corporate objectives in line with IE can also be achieved, separate from the improved environmental performance, by taking preventive actions not only with hard abilities but also with soft abilities.

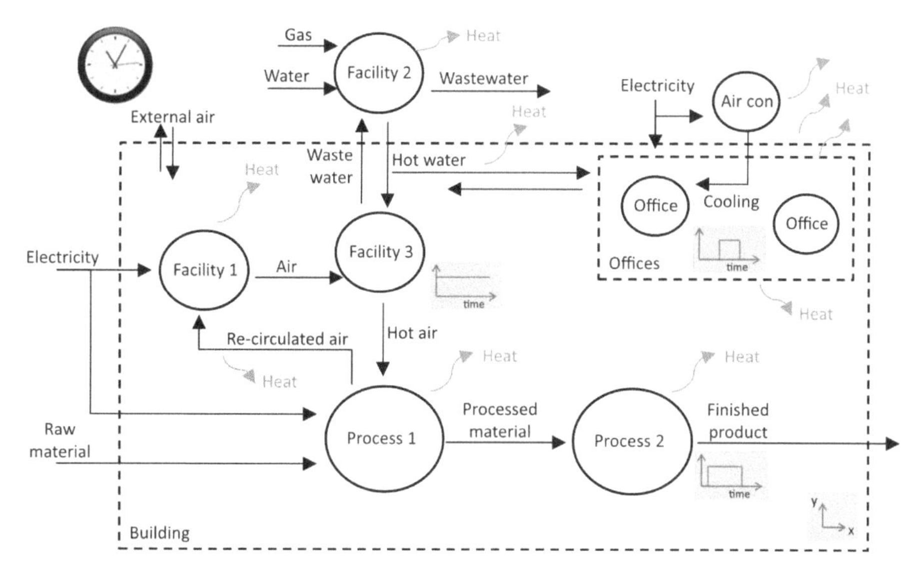

Figure 5. Manufacturing ecosystem example; snapshot at a specific time during operations.

Companies must make investments such as clean technologies, clean energy and raw materials with fewer contaminants. On the other hand, the soft abilities of an environmental cultural organization include pollution prevention, ISO 1400:2015, environmental lean thinking, environmental training, continuous environmental improvement and life cycle assessment (Avalos 2018).

The Staircase of Concepts

The staircase in Figure 6 considers all concepts in environmental management systems like ISO 14001:2015, so when a company tries to implement the system, it must be a strategic process and not an operative process. The final objective is to find a sustainable advantage over company rivals by reducing pollution in a preventive

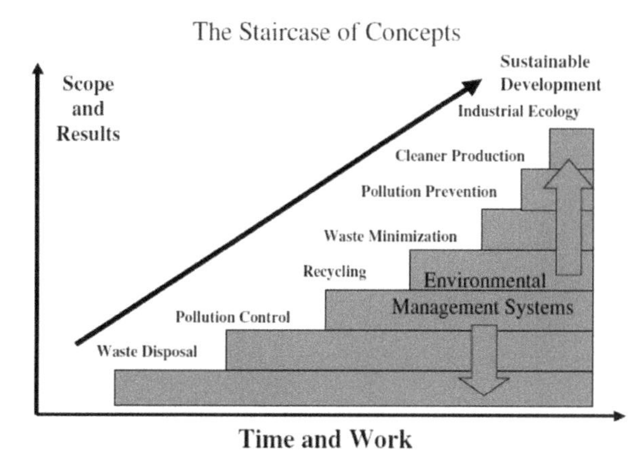

Figure 6. The staircase of concepts for environmental management systems (Hamner 1996).

manner. Waste disposal, pollution control and recycling must exist within the end-of-pipe technologies (corrective actions). Thus, these technologies would apply after preventive actions. Pollution prevention, cleaner production and IE must be applied as preventive responses not only for the influx of raw materials but also within production processes in order to reduce pollution from the source. IE as a corporate strategy is the driver for sustainable development. These preventive actions generate value for the company and deliver greater value for the customers (Avalos 2018).

The Environmental Management Programme

The relationship shown in Figure 7 provides a mechanism to apply local/operational activities at the global level using a bottom-up structure. This structure positions IE as a delivery tool for sustainable development goals. The sustainable development can be operationalized through an IE framework, which would be used for developing relevant cleaner production and P2 strategies at the corporate or organizational level for effective decision-making. Waste minimization and recycling are mainly planning activities at the operational level, which could be integrated into both long- and short-term planning. Pollution control and waste disposal are part of production activities at the operational level satisfying the strategies and policies of cleaner production and P2 using the IE framework at a top-level. P2 (Pollution Prevention Directive) means the use of materials, processes or practices that reduce or eliminate the creation of pollutants or wastes at the source. It includes practices that reduce the use of hazardous materials, energy, water or other resources and practices that protect natural resources through conservation or more efficient use (Heilala et al. 2008).

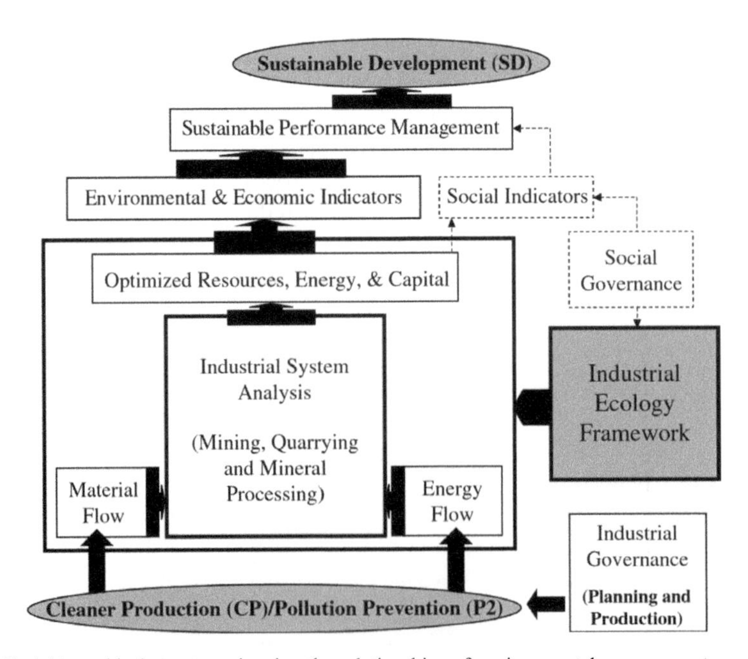

Figure 7. A hierarchical structure showing the relationships of environmental management programme concepts at various levels.

Industrial Ecology Framework Implementation in the Mining and Minerals Industry

"The concept of 'industrial metabolism', developed by Robert Ayres in 1989, is a system by which industry uses material and energy flow and then dissipates spent and unused products as wastes. Using a methodical approach of recording material and energy flow in an industrial system and by performing mass balances, one could identify inefficient processes or products and their negative impacts on an ecosystem.

Besides, an industrial system cannot be considered in isolation from its surroundings, which is the most important reason for implementing an IE framework at a mining or mineral processing operation. There are three factors of IE that must be optimized: resources, energy and capital. A major IE objective is to optimize these components" (Heilala et al. 2008).

As it shows on Figure 8; "Cleaner production and pollution prevention are used here interchangeable. Cleaner production, a subset of industrial ecology, promotes growth with minimal environmental impact under present technological limits. Cleaner production strategies should be developed at a corporate/organizational level, which would help develop action plans for its subsets: waste minimisation, recycling, pollution control, and waste disposal at a production level. Cleaner production is a dynamic process, which strives for continuous improvement" (Heilala et al. 2008).

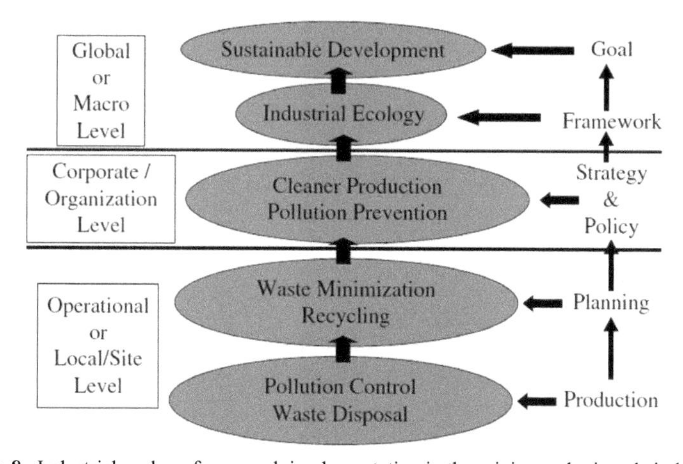

Figure 8. Industrial ecology framework implementation in the mining and minerals industries.

Cleaner Production

In 1987, the World Commission on Environment and Development first launched the term 'cleaner production' (CP; Hens et al. 2018). In 1990, UNEP defined CP as follows: "CP is the continuous application of an integrated preventive environmental strategy to processes, products, and services to increase efficiency and reduce risks to humans and the environment" (UNEP).

Figure 9 demonstrates the operational level activities' linkages to the corporate strategy of CP. In the middle level, it has identified CP as a tool in order to look for

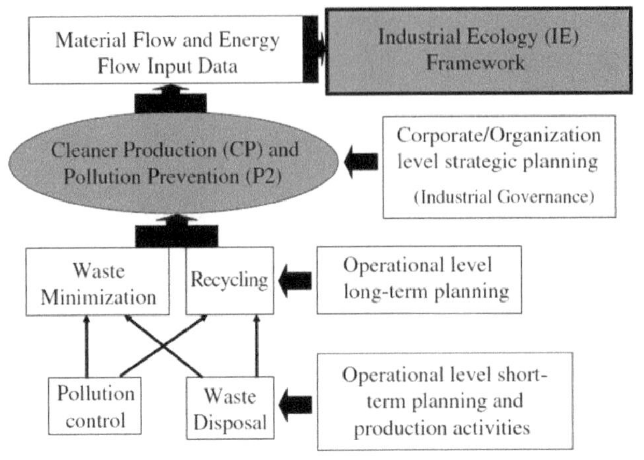

Figure 9. Cleaner production (CP) at the corporate and operational levels.

environmental improvement in the process production and make a greater contribution to IE. In other words, the CP tool is an effective driver with which to conduct sustainable management and IE. The CP generates environmental improvement on process production, making mass and energy balance on the production area, preventive maintenance, clean energy and clean technology. As a result, it will achieve an increase in productivity and strong waste reduction (Avalos 2018).

Accordingly, Cong and Shi (2019) stated that "Cleaner production is a key concept of industrial sustainable development as well as a research hotspot. In less-developed countries, cleaner production has incrementally emerged, and is especially concentrated on specialized industries." On the other hand, Zhang et al. (2015) argued that CP could be an excellent tool in order to develop sustainable projects and fulfill the stakeholders' requirements.

As an example of CP application, De Oliveira Neto et al. (2019) stated that "cleaner production practices brings as a result economic and environmental gains. The Sustainable Development Goals, reason why this research aims to evaluate if the economic and environmental advantages coming from cleaner production adoption in the textile industry contributed to the Sustainable Development Goals." Also says that: "it was concluded that the adoption of cleaner production practices in Brazilian textile industries through technological innovation made it possible to highlight the economic and environmental gains relating those to Sustainable Development Goals 9, 12, and 15."

The Tools for Implementing Industrial Ecology as Tactical and Operative Actions

Now it is going to describe and apply some tools that could help to continue with the implementation of the industrial ecology as a strategy in order to try to improve the environmental performance in a company, such as eco-design, environmental indicators, life cycle assessment, Sankey diagram, chemcad for the simulate process and structural model equation.

Eco-Design

A systematic integration of the environmental dimension early in the design process is very important for a significant reduction of environmental impacts. Considering environmental criteria in the same way as conventional design criteria is an objective the eco-design approach. Various methods and tools have been developed to enhance this integration and to provide designers with information regarding the environmental performance of products. Baumann et al. (2002) have identified eco-design methods and tools as supports to the design process, complemented by strategic and project organization tools. There are too main classes of tools to support the eco-design integration process: (1) non-specific tools to support the global coherence of the project; and (2) specific tools intended to perform the critical phases of the eco-design process, that is, the assessment and improvement of products environmental quality. Le Pochat et al. (2007) have refined the classification of "specific tools for eco-design into three categories according to their purpose: (1) environmental assessment of products, (2) environmental improvement and (3) both features at the time. Moreover, the observation of eco-design experts performing a redesign task pointed out three activities, which differentiate design from eco-design, namely: environmental assessment, solution finding and strategy definition" (Vallet et al. 2013).

"Even if the information flows is quite complete during the beginning of life phase (BOL, including design and production), it becomes unidirectional and non-continuous during the middle of life (MOL, including distribution, use, service and maintenance) and during the end of life (EOL, including reverse logistics, remanufacturing, reuse, recycling and disposal)" (Le Duigou et al. 2012, Demoly et al. 2013).

Table 1. A sample of eco-design guidelines for creating the improvement module.

Life cycle	**Strategy**	**Strategy deployment**
BOL: Extraction of materials production	Raw materials economy, Optimize production techniques	Use recycled materials, use materials with lower embedded energy. Reduce the number of production steps, and reduce energy consumption and waste generation in production. Valorize solid waste in production, reduce use of consumables.
MOL: Logistics and packaging	Reduce the impact of logistics Reduce the impacts of packaging	Choose fewer polluting means of transportation, reduce logistics distances. Reduce quantity, number, the weight of packaging and choose fewer impacting materials for packaging.
EOL	Reduce, Reuse, Recycle, (3R)/ Improve recycling Improve logistics	Favor reuse of product/components through design; encourage EOL product breakdown (e.g., reduce solids waste) and favor recycling of materials. Operate modal shift. Avoid empty returns.

Source: Adriankaja et al. (2015)

Embedded workflows

As is can see in Figure 10; the activities in the proposed eco-design method embed four phases of workflows:

Phase 1: Define de-scope, i.e., the project objective and the system boundary;
Phase 2: Frame the case parameter, set up the baseline and implement the initial environmental assessment;
Phase 3: Analyse the results, select the environmental improvement guidelines and generate new design scenarios, and
Phase 4: compare new design scenarios with the baseline (Adriankaja et al. 2015).

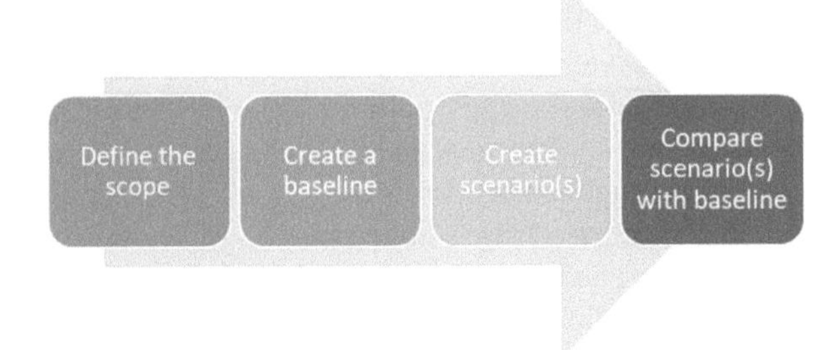

Figure 10. Eco-design method embedded workflows.

Indicator of Environmental Business Performance

It is very important to measure the environmental performance of an organization in order to find environmental improvement opportunities. According to Dragomir (2018), "in the last decades, the European Union has developed a set of environmental policies whose main objectives have been to protect natural capital and to develop a resource efficient and green economy. On his research, he found that Sweden, Austria, Denmark, Italy and Germany have the better scores in environmental performance assessment. However, Slovakia, Romania, Czech Republic, Poland and Lithuania have the lowest scores in the composite index." On the other hand, Trumpp et al. (2017) analyzed 11 scholarly definitions about corporate environmental performance and arrived at the conclusion that the ISO 14031 definition ('the measurable results of an organization's management of its environmental aspects') is the best definition available because it is encompassing and parsimonious and results from expert and academic consensus.

Dragomir, V.D. (2018) has proposed the following table.

Table 2 shows some indicators that could be used in different sectors in order to evaluate environmental performance and look for environmental improvement considering the maximum limit permissible (MLP) in any country. It is suggested to try to improve the environmental performance at a lower level with the objective of focussing on production but not environmental laws. As an example, if the MLP is

Table 2. An inventory of quantitative environmental performance indicators.

Dimensions	**Indicators**
Input (I) Materials Energy Water Company-wide water accounting Targets and initiatives	Water withdrawn by source Energy consumption by type and source, renewable or non-renewable Fossil fuel consumption: coal, oil, gas The energy intensity ratio for the organization Total land owned, leased or otherwise occupied by the company Percentage of terrestrial surface area that has rainfall collection network Percentage of inputs where the env. quality is controlled Primary and auxiliary materials used, renewable and non-renewable Reduction in total materials used Quantity of raw materials reused in the production process Substitution by renewable materials Use of recycled/waste materials Substitution by renewable energy source Substitution by alternative energy source Increase in co-generation facilities The rate of introduced chemicals with lower environmental impact
Output (O) Emissions Effluents and waste Targets and initiatives Emissions methodology/data Company-wide water accounting	Emissions to air: NOx, SO_2, CO_2, VOC, CO and NH_3 Emission permit Toxic releases: mercury, cadmium, chromium, lead GHG emission by scope e direct, indirect and products Emissions to water: COD, P, N, AOX and others Spillage in the soil Waste (non-hazardous, hazardous) generated and transported Waste per unit of production Wastewater Noise, heat, vibration, light, radiation Quantity of energy units saved due to energy conservation programs Dust emissions

Source: Dragomir (2018)

100 tonne/day for SO_2 emissions, the goal must be 50 tonne/day or less because the environmental rules change from time to time (Avalos 2018).

Denmark has been characterized by having values in ecolabel, licenses and environmental management systems that are more than twice the EU-28 averages (Eurostat). Moreover, in the category of efficient resource use, this member state has greater values in all of their components (resource productivity, eco-innovation index and efficient mobility of public transport) than the EU 28-average.

Life Cycle Assessment (LCA)

LCA considers the environmental aspects and impacts of services, products or processes. It is guided by ISO 14040 and ISO 14044 (ISO, 2006a, 2006b). According to those standards and LCA, the study must comprise four stages: goal and scope definition, inventory analysis, impact assessment and interpretation.

LCA is an excellent tool for a company that is looking for a better manner of improvement in environmental performance. This method could lead organizations to find the greatest impacts from the entire value chain in order to improve not only environmental performance but also the environmental quality of the product. LCA provides information to the companies on exchanging for friendly raw materials and redesigning the processes and products in order to deliver greater value to the customers. LCA is useful for the mining, cement, energy, chemical, petroleum, food, agriculture and automotive industries (Avalos 2018).

The companies must be ahead about the environmental regulation, and in that way LCA is an outstanding tool because the impact information that gives the LCA can be a source of innovation in order to redesign the process, product or services in order to be not only more competitive but also fulfill the environmental regulation. (Avalos 2018).

Let us cite an example of how LCA is applied on the cement industry:

"The construction sector generates several environmental problems and the use of sustainable building materials has become the main focus of research and development in achieving the goal of sustainable construction (Pacheco-Torgal et al. 2014). In this context, it is very important to understand the environmental impacts of cement production, since it is the main component of concrete, the most consumed material on Earth (WBCSD 2009a). The worldwide cement production reached 3.6 billion tonnes in 2012 (Cembureau 2012), and it is projected to grow by 0.8–1.2 percentage per year, reaching between 3700 and 4400 megatons in 2050 (WBCSD 2009a)."

"The key component of cement is clinker, which is a mixture of nodules and lumps of tri- and dicalcium silicates (alite and belite), tricalcium aluminate and tetracalciumaluminoferrite, which is produced by sintering of calcium oxide, aluminosilicates and other raw materials (Martos and Schoenberger 2014). This process is based on the decomposition of calcium carbonate into calcium oxide, which causes high carbon dioxide (CO_2) emissions, in addition to those associated with burning of fossil fuels. Figure 11 shows a typical flowchart of ordinary Portland cement production."

"Among those alternatives, the use of wastes as raw material and fuels is called co-processing and is well implemented in many places worldwide. The most common wastes are tyres, wood waste, plastics, meat and bone animal meal, municipal waste as refuse derived fuel (RDF), sewage sludge, and textiles. In the EU, the thermal substitution ratio in cement kilns increased from 3% in 1990 to 16.7% in 2004. In fact, countries like Austria, Germany and Norway reached substitution ratios above 60%. In 2010–2011, the Netherlands reached a replacement ratio of 83% while the average for EU was about 30%. Those numbers confirm the capacity of the cement industry for co-processing wastes" (Aranada Usón et al. 2013). In the research by

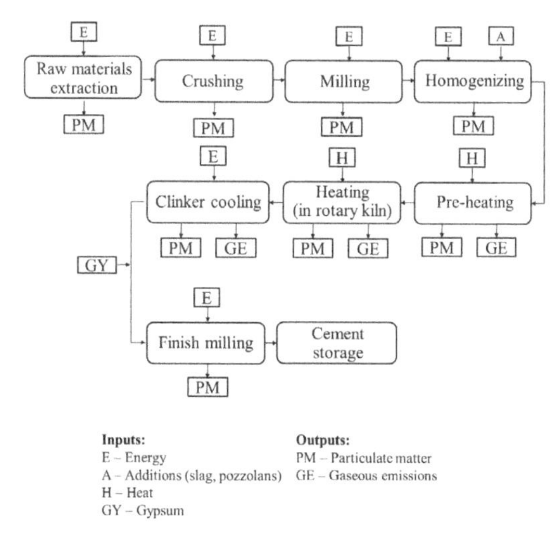

Figure 11. Process flow diagram for the manufacture of cement. Adapted from Huntzinger and Eatmon (2009).

Supino et al. (2016), the co-processing in cement industries has played a pivotal role, producing a triple win: emissions reduction, decreases in the extraction of natural resources and fossil fuels and enhancement of waste management operations.

Methodology

There is no single way to carry out LCA studies. ISO 14040 recognizes that organizations must have the flexibility to implement LCA in accordance with the intended application and their own requirements (ISO 14040, 2006). Nevertheless, some steps should be followed: (1) definition of the goal and scope, (2) inventory analysis, (3) impact assessment and (4) interpretation.

Goal and Scope

The goal was to assess the environmental impacts of using wastes as fuels in cement manufacturing based on primary data from process plants in southern Europe. A process flowchart is shown in Figure 12 in order to identify the main inputs and outputs concerning this activity. The functional unit is one tonne of ordinary Portland cement. Within the boundaries of the system are extraction and processing of raw materials and fossil fuels, alternative fuel supplies, all transport steps involved and the unit operations required to produce the cement at the manufacturing plant. Capital goods were included. RDF and scrap tires were used as alternative fuels partially replacing fossil fuels. Plastic and rubber, polyurethane foam and industrial wastes together represented less than 1% of the alternative fuel composition and were therefore not included in this study.

Inventory Analysis

The inventory data for the foreground system were taken directly from the studied industrial unit, located in Portugal, southern Europe and were based on data collected

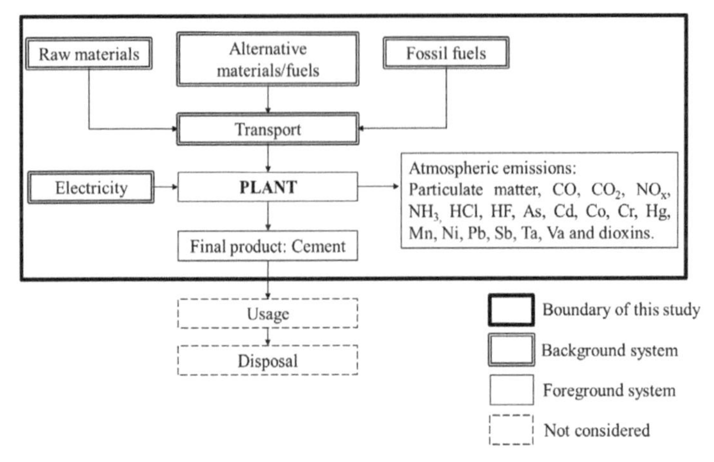

Figure 12. System boundaries of the LCA of a cement plant in southern Europe.

in 2013. Inventory data corresponding to the background system (raw material extraction, fossil fuel procurement and electricity production) were taken from the Ecoinvent database (Sauter et al. 2013).

Impact Assessment

Life cycle impact assessment was conducted using a database for characterization factors of life cycle impact assessment, which was developed by Centrum voor Milieuwetenschappen Leiden (CMD; Institute of Environmental Sciences 2013).

The evaluated impact potentials were the abiotic depletion potential (ADP), acidification potential (AP), eutrophication potential (EP), global warming potential (GWP) and photochemical oxidation potential (POP). These impact categories were chosen because the substances referred to in the inventory mostly affect them. Moreover, these five impact categories covered local, regional and global impacts from cement manufacture.

All calculations were performed using the LCA software Simapro (Prè Consultants 2014). In order to compare those results to other studies (Chen et al. 2010, Josa et al. 2007, 2004), the impact assessment was conducted based on the same impact assessment method (CML 2001, last updated in 2013).

Results

The impact assessment results for each step considered in cement manufacture are shown in Figure 13, and the absolute values for each impact category are presented in Table 3. The fossil fuel production step is the main contributor to ADP, but it is important to consider that petroleum coke is responsible for 99% of the measured impact potential. For the remaining impact categories, the atmospheric emissions from the cement kiln step have the biggest contribution. This was expected especially for GWP because of the CO_2 emissions associated with the calcination reaction.

$$CaCO_{3(s)} \rightarrow CaO_{(s)} + CO_{2(g)}.$$

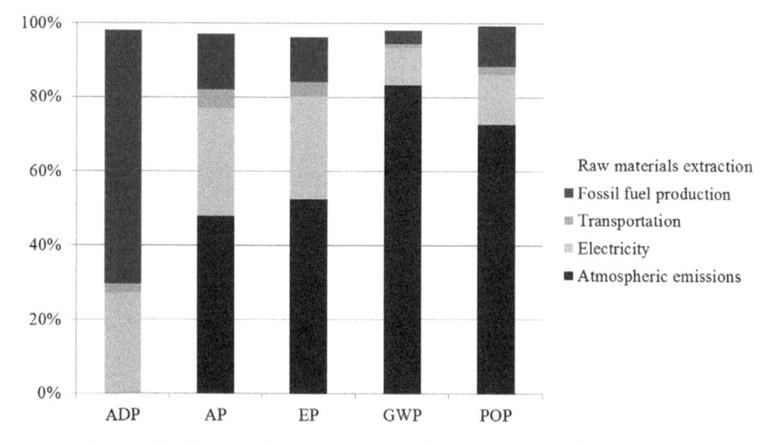

Figure 13. Contribution of each processing step to the studied impacts.

Table 3. Impact assessment of 1 tonne of cement according to CML 2001.

Impact	Unit	Total
ADP	{Kg Sb-Eq}	1.81×10^{0}
AP	{Kg SO_2-Eq}	1.971×10^{0}
EP	{Kg PO_4-Eq}	3.54×10^{-1}
GWP	{Kg CO_2-Eq}	$6.32 \times 10^{+2}$
POP	{Kg C_2H_4-Eq}	1.58×10^{-1}

Source: (Stafford 2016)

This reaction is responsible for approximately 60% of the CO_2 emissions from the kilns, which corresponds to 83.2% of CO_2 equivalent emitted by this system. GWP is the impact category most discussed in cement production studies.

Concerning AP, most of the total impact is derived from the kiln emissions, which correspond to 48% of the total. In this specific case, most of the SO_2 emissions arise from the kiln because of fuel combustion and the processing of the raw material in the kilns (Pacheco-Torgal et al. 2014). In fact, SO_2 production results from the oxidation of sulfide or elemental sulfur contained in the fuel or raw materials when there are sufficient oxygen and the material temperature is in the range of 300°C to 600°C (Basel Convention 2011).

From Figure 13, it is possible to verify that EP causes are very similar to AP causes. The atmospheric emissions from the kiln contribute to 52.6% of EP, and electricity and fossil fuel use correspond to 27.7% and 12.3%, respectively. The last analyzed category, POP is mostly influenced by atmospheric emissions (72.6%), followed by electricity (13.6%) and fossil fuel use (10.9%). As those last three categories are influenced by nitrogen and sulfur compounds, it is natural that the atmospheric emissions from the kiln contribute the most. Besides this, carbon compounds also influence POP, which is present in these emissions.

According to Chen et al. (2010), the atmospheric emissions from the kiln are responsible for 88.6% of GWP, and in this case the value is 83.2%. They also took into account the use of wastes as alternative fuels, but this procedure was not detailed. Thus, we cannot affirm that the lower values were found to occur because of the RDF usage.

In this instance, we have seen that LCA is a very useful method for finding opportunities for environmental improvement. Considering this LCA for the cement industry, it is very clear that the focus must be on the reactor where the raw materials are treated as it generates the most particulate matter and atmospheric emissions. In the future, the objective of that industry must be zero-emissions (Avalos 2018).

Sankey Diagram

A Sankey diagram is used to represent many flow visualizations in order to show how the resources of an organization are transformed, starting with the raw materials and finishing with obtaining the product. The magnitude of the arrow line represents the quantity of the flow. The data input is the mass or energy balance of any process in order to identify the efficiency of the transformation processes. The Sankey diagram can also show the pollution of the process, identifying atmospheric emissions, solid wastes and effluents.

For instance, the Sankey diagram for the copper industry is shown in Figure 14 (it is not to scale). The diagram shows that for the production of one tonne of fine copper, 150 tonnes of ore material must be processed. In addition, 2.8 tonnes of particulate matter and atmospheric emissions (sulfur dioxide), 1.8 tonnes of slags and 145 tonnes of tailing ponds are generated, where the most critical environmental problem is the SO_2 emission. For the SO_2 emissions, one alternative could be the transformation of SO_2 into sulfuric acid. The tailing ponds could be transferred to a secure landfill with the slags separated first; the solid wastes must be transferred to a secure landfill, and the wastewater must be sent to an industrial water treatment plant.

Sankey diagram is very useful when a company is looking for a critical environmental improvement because, as can be seen for this case (copper industry); the most critical environmental problems are the SO_2 emissions and particulate matter.

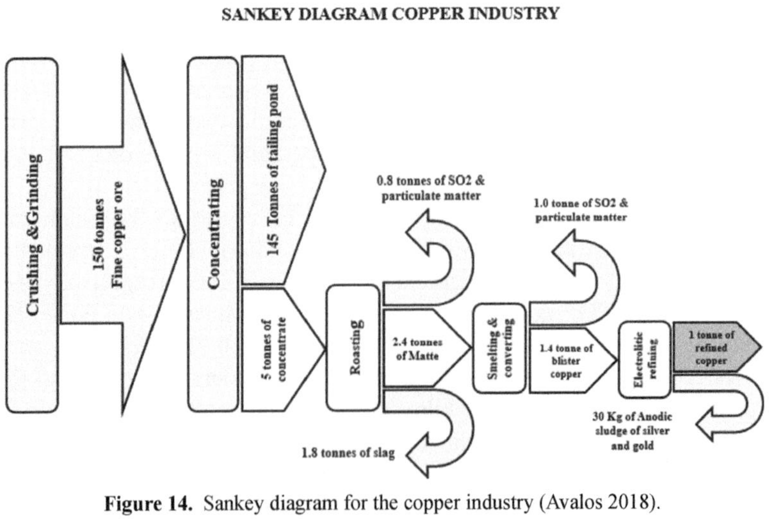

SANKEY DIAGRAM COPPER INDUSTRY

Crushing &Grinding
150 tonnes Fine copper ore
Concentrating
145 Tonnes of tailing pond
5 tonnes of concentrate
Roasting
0.8 tonnes of SO2 & particulate matter
2.4 tonnes of Matte
1.8 tonnes of slag
Smelting & converting
1.0 tonne of SO2 & particulate matter
1.4 tonne of blister copper
Electrolitic refining
1 tonne of refined copper
30 Kg of Anodic sludge of silver and gold

Figure 14. Sankey diagram for the copper industry (Avalos 2018).

Process Simulation with ChemCad

In recent years, the improvement of the existing chemical processes, the so-called retrofit, has become increasingly important because of escalating prices for energy and feedstock. Thus, there is a growing market for the design and improvement of highly efficient processes. The corresponding increase in computational performance has led to the use of mathematical methods to address these challenges (Zobel et al. 2006). The development of new processes is carried out using highly efficient process synthesis methods under consideration of efficiency and sustainability criteria (Halim et al. 2011). These mathematical approaches use powerful algorithms and programming software. The advantages of real plant optimization include an analysis and subsequent improvement of plant performance, the ability to obtain information about the influences of production changes and the detection of bottlenecks and problems. Kiss et al. (2010) successfully showed the advantages of modeling a real plant with a commercial simulator, fitting the simulation to the real plant data and improving the performance with optimization. Effective process analysis and optimization tool are crucial for the improvement and continuous development of existing plants.

Dynamic optimization methods can be used for the improvement of processes, especially if the desired process has to be flexible in terms of product changes or energy market prices or if the start-up procedure is complicated (Reepmeyer et al. 2004). The researchers reported a high number of function evaluations as the algorithm used resulted in an optimization horizon of up to several days. Zobel et al. (2005) extended this framework to the flowsheet simulator ChemCad and provided proof of the advantages of such a connexion, such as robustness.

Here is an example of a simulation with ChemCad 7.1.2.

Steady-State Simulation

A stream of 100 kmol/h of a mixture of ethanol (50% mol) and n-propanol (50% mol) is fed to a continuous distillation column at ambient temperature (25°C) and atmospheric pressure. Pressure drop through the column is negligible, and a reflux ratio of R = 1.5 Rmin is used. Approximately 93% of ethanol and 5% of n-propanol are desired to be in the distillate stream.

Design this process in ChemCad using a shortcut column to obtain the number of total stages, a minimum number of stages, location of feed stream, minimum and calculated reflux ratio, compositions of distillate and bottom fractions and energy requirements (heat duty) of the reboiler and condenser.

The basic procedure for creating a simulation can be broken down into the following common steps:

1. Start a new simulation.
2. Select engineering units for the simulation.
3. Create a flowsheet with the appropriate streams and unit operations.
4. Select chemical components for the process.
5. Select K-value and enthalpy options for the process.

6. Define the feed streams used in the process.
7. Enter specifications for the unit operations.
8. Run the simulation.
9. Review the results of the simulation.

As you can see on Figure 15; select Format/Engineering Units and then choose original or alternative SI units or change if you need it.

As you can see on Figure 16, Start by selecting a shortcut column from the UnitOp Palette on the right.

For some UnitOps, alternatives are shown as a dropdown menu by clicking the lower-left corner from the UnitOp.

Place the shortcut column on the main screen.

Complete the diagram with inputs, outputs and connectors.

Press Esc to exit draw mode. Then select chemical components for the process (As it can see on Figure 17).

Select K-value and enthalpy options for the process (As it can see on Figure 18).

Define the feed streams used in the process (As it can see on Figure 19).

Enter specifications for the unit operations (As it can see on Figure 20).

Run the simulation.

Run the simulation from the menu or with the green arrow button (As it can see on Figure 21).

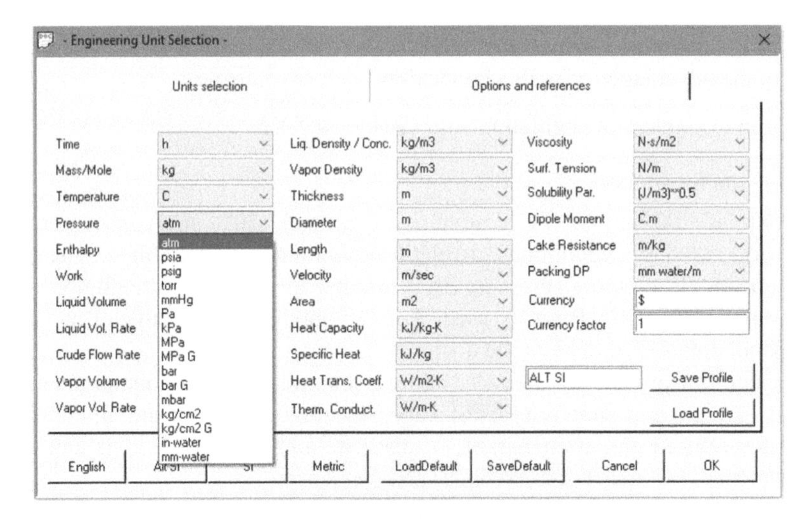

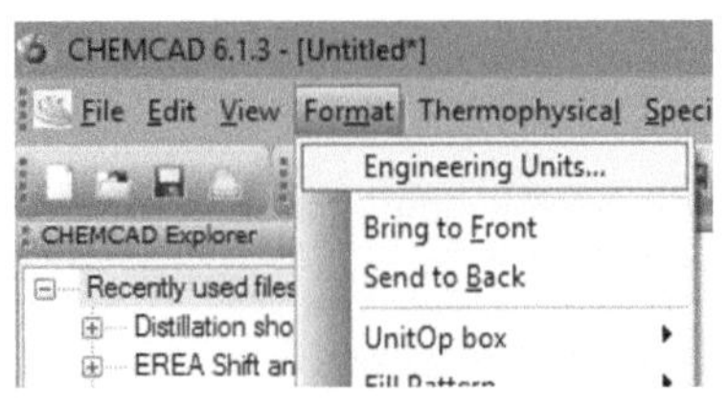

Figure 15. Select format and engineering units (Avalos 2018).

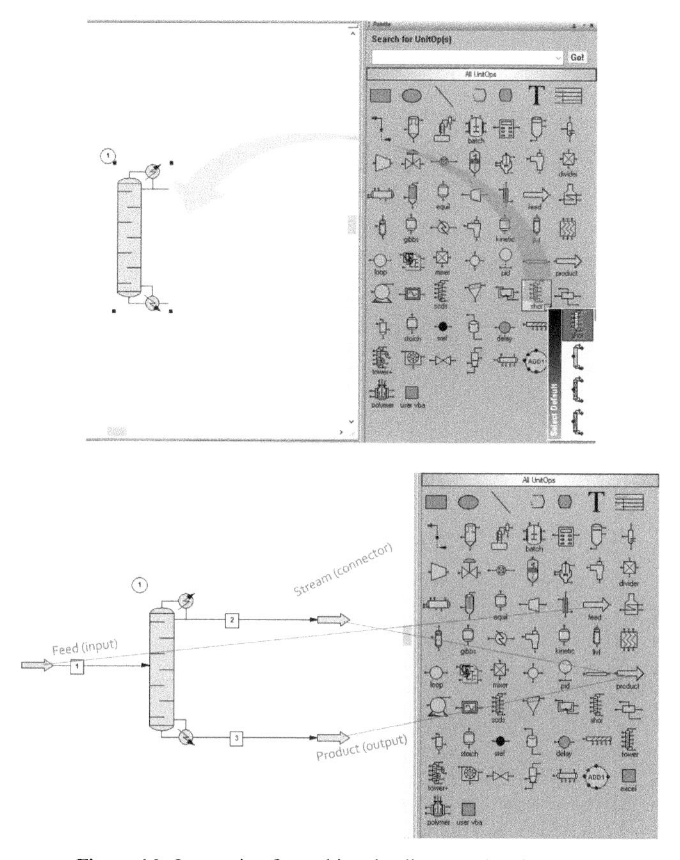

Figure 16. Instruction for making the diagrams (Avalos 2018).

- Menu: Thermophysical/select Components...
- Find and select components from the list (use name, CAS number or formula)
- Add components to the right window and press Ok

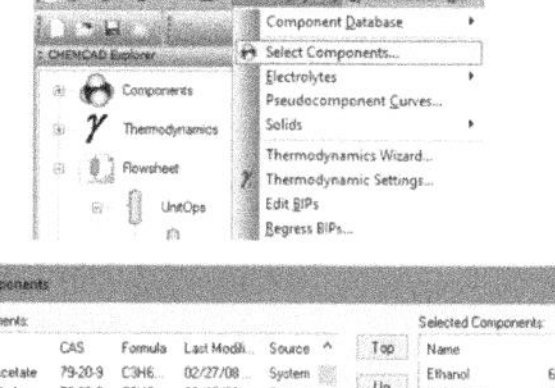

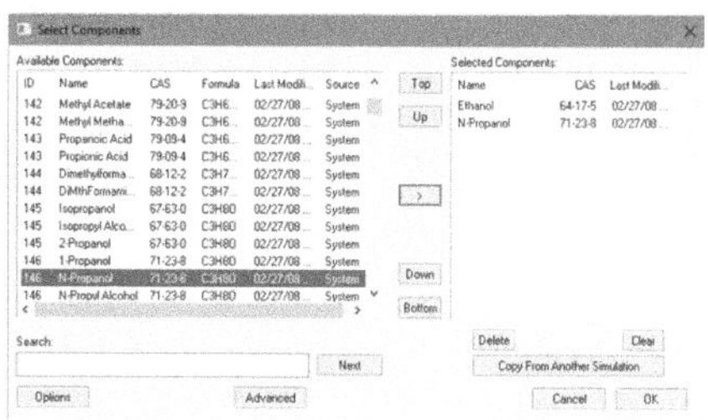

Figure 17. Select the chemical components (Avalos 2018).

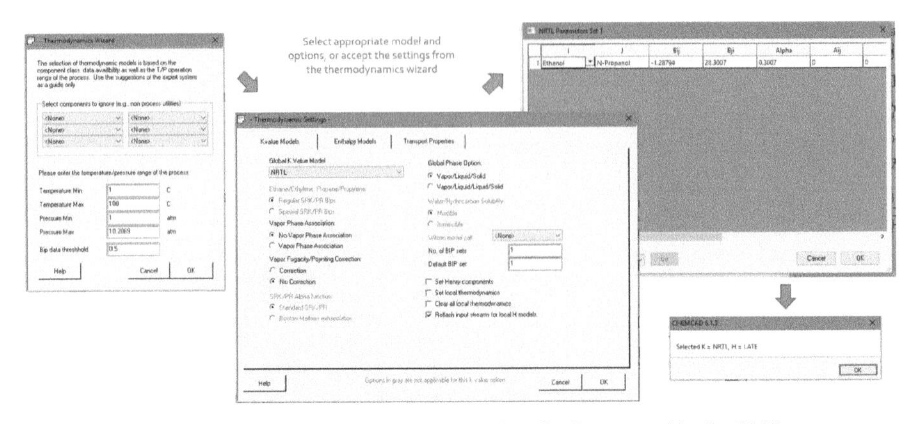

Figure 18. Select K-value and enthalpy options for the process (Avalos 2018).

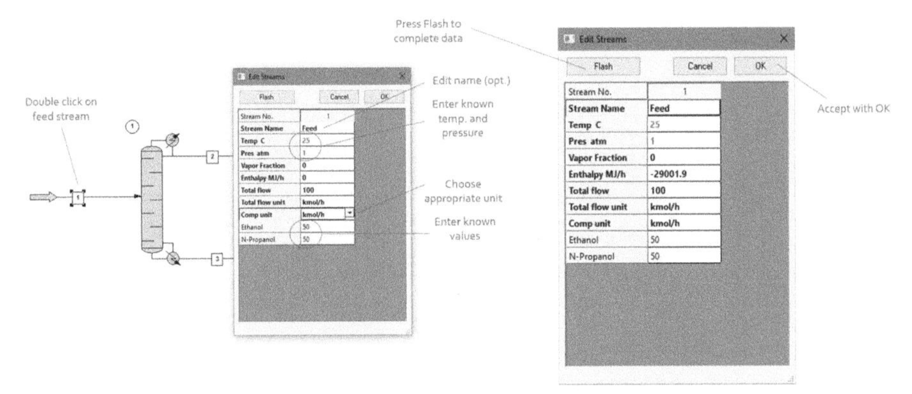

Figure 19. Define the feed streams used in the process (Avalos 2018).

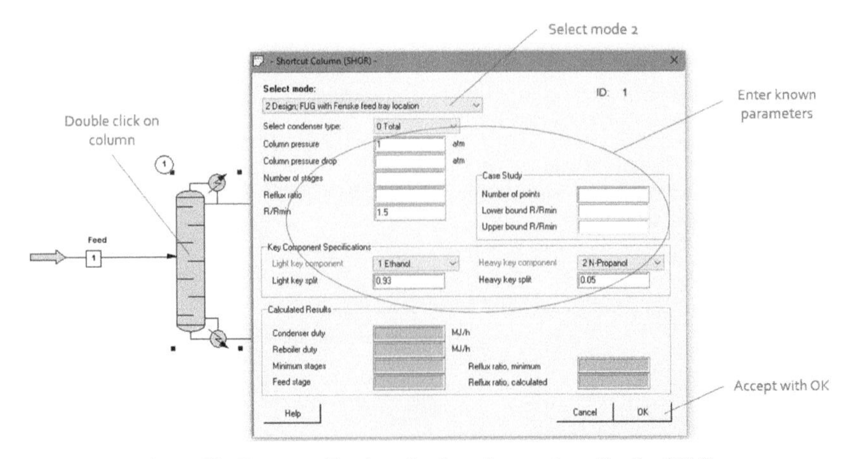

Figure 20. Enter specifications for the unit operations (Avalos 2018).

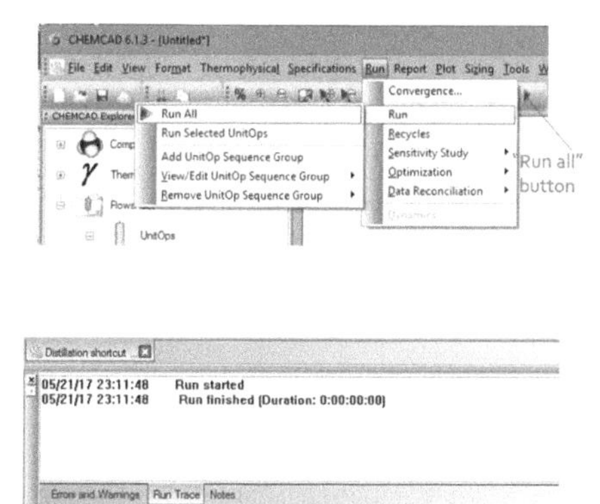

Figure 21. Run the simulation (Avalos 2018).

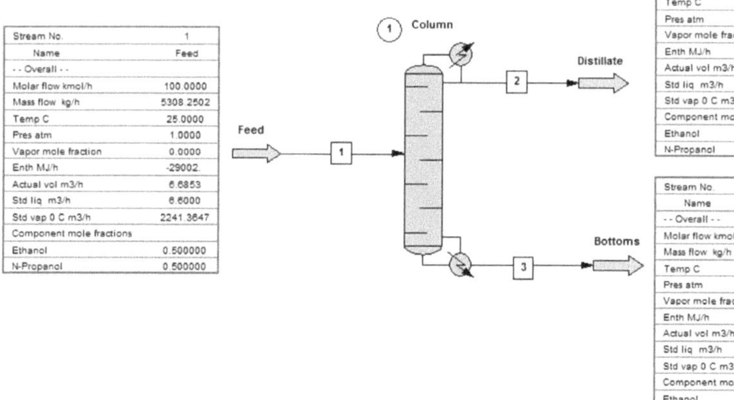

Figure 22. Result of the problem after running the model with ChemCad.

Check the notification panel at the bottom part of the screen.
If errors are reported, check all previous steps and data entered.
Review the results of the simulation (As it can see on Figure 22).

If you click on 'report' and 'mass' and 'energy balance', you obtain the following report:

Mass Balance Report

Overall Mass Balance	kmol/h		kg/h	
	Input	Output	Input	Output
Ethanol	50.000	50.000	2303.450	2303.449
1-Propanol	50.000	50.000	3004.800	3004.800
Total	100.000	100.000	5308.250	5308.250

Energy Balance

Overall Energy Balance	**MJ/h**	
	Input	**Output**
Feed Streams	–29001.9	
Product Streams		–28080.1
Total Heating	6905.13	
Total Cooling	–5983.36	
Power Added	0	
Power Generated	0	
Total	–28080.1	–28080.1

Mathematical Model with Structural Equations

"A structural equation modelling is used to test any theoretical model. Structural equation modelling is a statistical methodology that takes a confirmatory (i.e., hypothesis-testing) approach to the analysis of a structural theory bearing on some phenomenon" (Byrne 2006).

"Structural Equation Modelling (SEM) has become one of the techniques of choice for researchers across disciplines and increasingly is a 'must' for researchers in the social sciences. However, the issue of how the model that best represents the data reflects underlying theory, known as model fit is by no means agreed. With the abundance of fit indices available to the researcher and the wide disparity in agreement on not only which indices to report but also what the cut-offs for various indices actually are, it is possible that researchers can become overwhelmed by the conflicting information available" (Yuan and Bentler 2005).

For instance, Park (2019) researched 'Social Acceptance of Green Electricity' in South Korea and used SEM in order to explore both motivations and hindrances of the social acceptance of green electricity by users. This survey was conducted to test a research model that used the theory of planned behavior and the general model of perceived value with factors extracted from in-depth interview sessions. The structural results indicated that users' perceived value of green electricity is significantly affected by perceived benefits.

Extensive research has examined the relationship between environmental management/performance and business performance, but it yielded no conclusive results (Zeng et al. 2010).

As it can see on Figure 23, the author studied the copper industry in Perú using SEM and evaluated the hypothesis if there is a relationship between CP and business performance (Avalos 2018). The method and results will be presented in the sections that follow.

Method

Low- and high-cost activities have been established for CP activities, and financial and non-financial performances have been defined for business performance. In addition, 17 observable variables have been defined, which are specified in

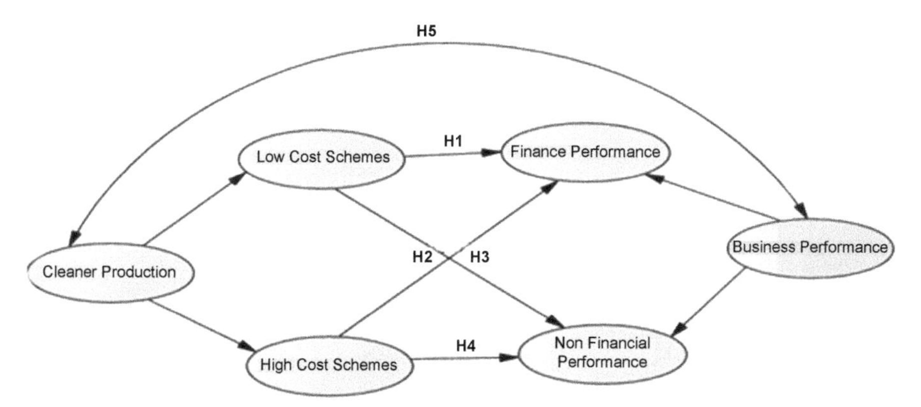

Figure 23. Relationship between cleaner production and business performance. Adapted from Zeng et al. (2010).

Table 3. The hexogen construct is CP, and the endogen construct is business performance. Twenty-one companies in Perú have been surveyed. The data were collected through a questionnaire survey. First, a pilot survey was conducted and then five experts were invited to evaluate the questionnaires. The final questionnaires were sent to 10 managers (from operation, maintenance, finance, environmental areas and strategic planning) of each company, for a total of 210 surveys collected. Ninety percent of the surveys collected had the questionnaire completed (the questionnaire comprised 53 questions). The data were run with SPSS 22.0 to evaluate the reliability analysis. The values of the Cronbach's alpha for the low- and high-cost schemes were 0.827 and 0.74, respectively, and for financial and non-financial performances, the resulting values were 0.854 and 0.795, respectively. Additionally, validity refers to the extent to which items reveal true information. The validity analysis included content validity and construct validity, the results of which are shown in Table 4.

Hypotheses

H_1: CP of the low-cost scheme has a positive impact on financial performance.
H_2: CP of the low-cost scheme has a positive impact on non-financial performance.
H_3: CP of the high-cost scheme has a positive impact on financial performance.
H_4: CP of the high-cost scheme has a positive impact on non-financial performance.
H_5: Insights into Hypotheses 1–4 provide us with a basis to make inferences on the relationship between CP and business performance. This can be seen in Figure 23.

Results

According to Table 5, it can see that the low-cost scheme has been found to have moderate contributions of 0.45 and 0.52 to financial and non-financial performances, respectively, as it can see in Figure 24. The high-cost scheme has strong contributions of 0.72 and 0.81 to financial and non-financial performances, respectively. In general, there is a strong relationship between CP and business performance: 0.83. Thus, companies should focus on both low- and high-cost schemes in order to improve their environmental performance.

Table 4. Indexes of cleaner production and business performance.

Indexes for evaluation	Observable variables	Code	Indicators evaluation	Observable variables	Code
Low-cost scheme	1. Training	LCCP1	Financial indicators	12. Profitability	FBP1
	2. Work conditions	LCCP2		13. Increase net of rate profit	FBP2
	3. Standardize procedure (environmental system)	LCCP3		14. Return on equity	FBP3
	4. Pollution prevention culture (strategic planning)	LCCP4	No financial indicators	15. Markey share	NFBP1
	5. Continuous process improvement (environmental system)	LCCP5		16. Corporate reputation	NFBP2
	6. Environmental management system (ISO 14001:2015) and life cycle assessment (ISO 14020, 14021).	LCCP6		17. Environmental performance	NFBP3
High-cost scheme	7. Clean technology (Isasmelt)	HCCP1			
	8. Clean energy	HCCP2			
	9. Renewable resources	HCCP3			
	10. Eco design products	HCCP4			
	11. Industrial ecology	HCCP5			

Source: (Avalos 2018)

Table 5. Results of the hypothesis test.

Hypothesis	Path relationship	Estándar path Coefficient (μ)	P Value	Result
H_1	Low-cost scheme® Financial performance	**0.45**	< 0.001	Support
H_2	Low-cost scheme® Non-financial performance	**0.52**	< 0.001	Support
H_3	High-cost scheme® Financial performance	**0.72**	< 0.001	Support
H_4	High-cost scheme® Non-financial performance	**0.81**	< 0.001	Support
H_5	Cleaner production® Business performance	**0.83**	< 0.001	Support

Source: (Avalos 2018)

The low-cost scheme was found to have a moderate contribution because its soft abilities help in improving financial performance by eco-efficient resource management and in improving non-financial performance (for instance, environmental performance) by the philosophy of continuous improvement and active participation on environmental management systems (ISO 14001, 14031). However, these alone are not enough.

The CP activities of the high-cost scheme have a greater contribution to both financial and non-financial performances. This is because investments in clean technology, clean energy and raw material with less contamination, eco-design,

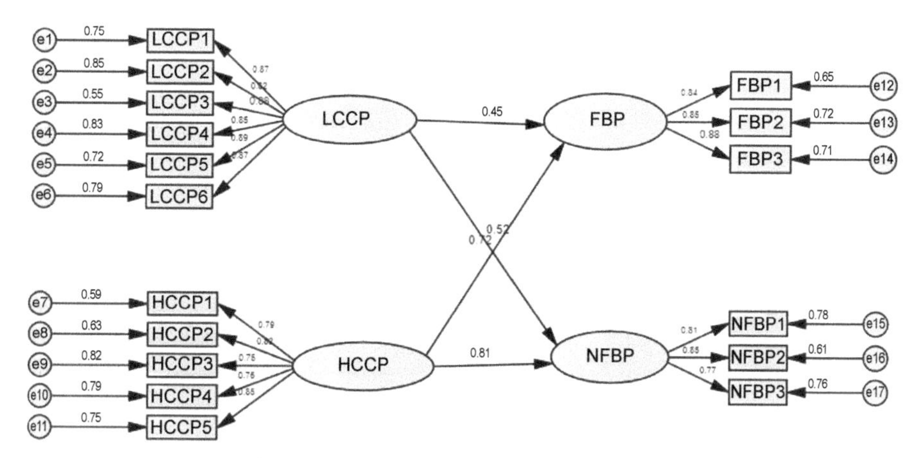

Figure 24. Path coefficients in SEM (Avalos 2018).

industrial ecology and life cycle assessment give a company the chance to improve processes where raw material transformation occurs. As a result, this greatly reduces wastes, such as effluents, atmospheric emissions and solid wastes, and contributes more heavily to environmental performance.

Conclusions

This chapter analyzed the question if there is a strong relationship between IE as a strategy and environmental performance. Regarding IE as a strategy, it was important to first note that traditional strategies like differentiation and cost leadership only focus on the economical result and do not consider environmental and social performance. Within the context that the organization is indeed sustainable, IE as a strategy can be an excellent alternative because it has attributes of both differentiation and cost strategy, like innovation and efficient resources management, but it also utilizes eco-efficient resource management, product eco-design and lean environment production, focussing on environmental aspects and resolving environmental pollution with preventive tools.

ChemCad has demonstrated that running a process simulation could feasibly find the best environmental performance for any kind of process production while minimizing costs and obtaining the best result. Using structural equation modeling, it has demonstrated that there is a strong relationship between CP and business performance. The value of 0.83 obtained suggests that all investments in soft and hard abilities will help a company achieve an excellent business result socially, economically and environmentally (for instance, investment in clean energy and technology).

Operational tools such as LCA, eco labels, CP, the Sankey diagram, green technology and green chemistry work as drivers of IE. The objective is to identify the most significant environmental problems within production and then try to solve these problems by using clean technology and mitigating pollution by purchasing raw materials with fewer contaminants.

Recommendations

It is suggested that companies adopt IE as a strategy in order to achieve more profitable and sustainable results, consisting of environmental, social and economic performance. After accepting IE as a strategy, a company should then implement life cycle assessment ISO 14040 in order to identify the most critical impacts along the value chain. Next, a company should make a Sankey diagram, beginning with the raw materials and ending with the product with the objective of identifying the most critical environmental impacts. It would be very useful to make a simulation using ChemCad in order to test the best parameters for the most critical environmental impacts, balancing mass and energy and creating sensibility studies for the variables. It also suggested that a company should study process production using SEM and by considering the tools ISO 14001, ISO 14030, 14040, Sankey diagram and eco-design as exogenous variables to prevent pollution and as endogen variables for business performance.

The objective is to determine the relationship between environmental preventive tools and business performance (both financial and non-quantitative performances). With these results, the company should implement an environmental management system ISO 14001:2015, which should be included in the company's strategic planning. IE as a strategy could help all manufacturing or transformation sectors (mining, cement, energy petroleum, food, plastic, etc.) achieve superior environmental and economical results (Avalos 2018).

Acknowledgments

I would like to acknowledge Dr. Berta Diaz, Dean of Architecture and Engineering Faculty and Dr. María Teresa Quiroz, Director of Research Institute IDIC from Universidad de Lima for sponsoring this work. I would also like to thank Dr. George Power for his great suggestions.

References

Avalos, E. 2018. Environmental improvement model for cleaner copper production and its influence on business management with structural equations. Industrial Data, 21. Universidad Nacional Mayor de San Marcos. http://dx.doi.org/10.15381/idata.v21i2.15604.

Adriankaja, H., Vallet, F., Le Duigou, J. and Eynard, B. 2015. A method to eco-design structural parts in the transport sector based on product life cycle management. Journal of Cleaner Production, 94, 165–76.

Basu, A. J. and Van Zyl, D. J. A. 2006. Industrial ecology framework for achieving cleaner production in the mining and minerals industry. Journal of Cleaner Production, 14(3–4), 299–304. https://doi.org/10.1016/j.jclepro.2004.10.008.

Byrne, B. M. 2006. Structural Equation Modeling with EQS: Basic Concepts, Applications, and Programming (2nd ed.). Lawrence Erlbaum Associates Publishers.

Cong, W. and Shi, L. 2019. Heterogeneity of industrial development and evolution of cleaner production: Bibliometric analysis based on JCLP. Journal of Cleaner Production, 212, 822–836. https://doi.org/10.1016/j.jclepro.2018.12.024.

Dessus, S. and Bussolo, M. 1998. Is there a trade-off between trade liberalization and pollution abatement? Journal of Policy Modelling, 20, 11–31. https://doi.org/10.1016/S0161-8938(96)00092-0.

Despeisse, M., Ball, P. D., Evans, S. and Levers, A. 2012. Industrial at factory level—a conceptual model. Journal of Cleaner Production, 31, 30–9.

Despeisse, M., Oates, M. R. and Ball, P. D. 2013. Sustainable manufacturing tactics and cross-functional factory modelling. Journal of Cleaner Production, 42, 31–41. https://doi.org/10.1016/j.jclepro.2012.11.008.

De Oliveira Neto, G. C., Ferreira Correia, J. M., Silva, P. C., de Oliveira Sanches, A. G. and Lucato, W. C. 2019. Cleaner production in the textile industry and its relationship to sustainable development goals. Journal of Cleaner Production, 228, 1514–1525. https://doi.org/10.1016/j.jclepro.2019.04.334.

Diao, X. D., Zeng, S. X., Tam, C. M. and Vivian, W. Y. 2009. EKC analysis for studying economic growth and environmental quality: a case study in China. Journal of Cleaner Production, 17, 541–548.

Dragomir, V. D. 2018. How do we measure corporate environmental performance? A critical review. Journal of Cleaner Production, 196, 1124–1157. https://doi.org/10.1016/j.jclepro.2018.06.014.

Figueroa, B. E., Orihuela, R. C. and Calfucura, T. E. 2010. Green accounting and sustainability of the Peruvian metal mining sector. Resources Policy, 35, 156–167. https://doi.org/10.1016/j.resourpol.2010.02.001.

García-Álvarez, M. T. and Moreno, B. 2018. Environmental performance assessment in the EU: A challenge for the sustainability. Journal of Cleaner Production, 205, 266–280. https://doi.org/10.1016/j.jclepro.2018.08.284.

Graedel, T. E. and Allenby, B. R. 2003. Industrial Ecology. 2nd ed. New Jersey: Prentice Hall.

Hair, J., Anderson, R., Tathan, R. and Black, W. 2007. Modelos de ecuaciones estructurales. Análisis Multivariante, 491, 603–663. 6 ed. Madrid: Prentice Hall.

Halim, R., Gladman, B., Danquah, M. K. and Webley, P. A. 2011. Oil extraction from microalgae for biodiesel production. Bioresource Technology, Jan; 102(1), 178–85. doi: 10.1016/j.biortech.2010.06.136. Epub 2010 Jul 23. PMID: 20655746.

Hamner, B. 1996. What is the relationship between cleaner production pollution prevention, waste minimisation and ISO 14001? Asian Conf. Clean. Prod. Chem. Ind.

Hart, S. L. 1995. A natural-resource-based view of the firm. The Academy of Management Review, 20, 986–1014.

Hart, S. and Ahuja, G. 1996. Does it pay to be green? An empirical examination of the relationship between emission reduction and performance. https://doi.org/10.1002/(SICI)1099-0836(199603)5:1<30:: AID-BSE38>3.0.CO;2-Q.

Heilala, J., Vatanen, S., Tonteri, H., Montonen, J., Lind, S., Johansson, B. and Stahre, J. 2008. Simulation-based sustainable manufacturing system design. Winter Simulation Conference Proceedings (WSC'08), 7–10 December 2008, Miami, FL (United States), pp. 1922–1930.

Hens, L., Block, C., Cabello-Eras, J. J., Sagastume-Gutierez, A., Garcia-Lorenzo, D., Chamorro, C., Herrera Mendoza, K., Haeseldonckx, D. and Vandecasteele, C. 2018. On the evolution of "Cleaner Production" as a concept and a practice. Journal of Cleaner Production, 172, 3323–3333. https://doi.org/10.1016/j.jclepro.2017.11.082.

Hussey, D. 2007. Using structural equation modelling to test environmental performance in small and medium-sized manufacturers: can SEM help SMEs? Journal of Cleaner Production, 15, 303–312.

Jia, X. and Liu, L. 2008. Comparing the structural equation model with the simultaneous equation model. Applied Statistics Management, 27, 439–446.

José F. Molina-Azorín, Enrique Claver-Cortés, Maria D. López-Gamero and Juan J. Tarí. 2009. Green management and financial performance: a literature review. Management Decision, 47(7): 1080–1100. https://doi.org/10.1108/00251740910978313.

Kiss, A. A., Bildea, C. S. and Grievink, J. 2010. Dynamic modeling and process optimization of an industrial sulfuric acid plant. Chemical Engineering Journal, 158(2), 241–9.

López-Gamero, M. D., Claver-Cortés, E. and Molina-Azorín, J. F. 2009. Complementary resources and capabilities for an ethical and environmental management: a qual/quan study. Journal of Business Ethics, 82, 701–732.

Molina-Azorín, J. F., Tarí, J. J., Claver-Cortés, E. and López-Gamero, M. D. 2009. Quality management, environmental management and firm performance: a review of empirical studies and issues of integration. International Journal of Management Review, 11, 197–222.

Montabon, F., Sroufe, R. and Narasimhan, R. 2007. An examination of corporate reporting, environmental management practices and firm performance. Journal of Operations Management, 25(5), 998–1014. https://doi.org/10.1016/j.jom.2006.10.003.

Otte, D., Lorenz, H. M. and Repke, J. U. 2015. A toolbox using the stochastic optimisation algorithm MIPT and Chemcad for the systematic process retrofit of complex chemical process. Comp. Chem. Engn., 84, 371–81.

Padash, A. and Ghatari, A. R. 2020. Toward an innovative green strategic formulation methodology: empowerment of corporate social, health, safety and environment. Journal of Cleaner Production, 261, 121075. https://doi.org/10.1016/j.jclepro.2020.121075.

Park, E. 2019. Social acceptance of green electricity: Evidence from the structural equation modelling method. Journal of Cleaner Production, 215, 796–805. https://doi.org/10.1016/j.jclepro.2019.01.075.

Porter, M. E. and Van Der Linde, C. 1995. Green and competitive: ending the stalemate. Harvard Business Review, 28, 128–129. https://doi.org/10.1016/0024-6301(95)99997-E.

Porter, M. 1996. What is strategy? Harvard Business Review, 74, 61–78.

Reepmeyer, F., Repke, J. and Wozny, G. 2004. Time optimal star-up strategies for reactive distillation columns. Chemical Engeenering Sciences, 59(20), 4339–47.

Russo, M. V. and Fouts, P. A. 1997. A resource-based perspective on corporate environmental performance and profitability. The Academy of Management Journal, 40, 534–559. https://doi. org/10.2307/25705.

Salomons, W. 1995. Environmental impact of metals derived from mining activities: Process, prediction, prevention. Journal of Geochemical Exploration, 52, 5–23.

Sauter, P., Witt, J., Billig, E. and Thrän, D. 2013. Impact of the renewable energy sources act in Germany on electricity produced with solid biofuels – Lessons learned by monitoring the market development. Biomass and Bioenergy, 53(1), 162–171. https://doi.org/10.1016/j.biombioe.2013.01.014.

Schaltegger, S. and Synnestvedt, T. 2002. The link between "green" and economic success: environmental management as the crucial trigger between environmental and economic performance. Journal of Environmental Management, 65, 339–346.

Stafford, F., Dias, A. C., Arroja, L., Labrincha, J. A. and Hotza, D. 2016. Life cycle assessment of the production of Portland cement: a Southern Europe case study. Journal of Cleaner Production, 126, 159–165.

Supino, S., Malandrino, O., Testa, M. and Sica, D. 2016. Sustainability in the EU cement industry: The Italian and German experiences. Journal of Cleaner Production, 112, 430–442. https://doi.org/10.1016/j.jclepro.2015.09.022.

Testa, F., Rizzi, F., Daddi, T., Gusmerotti, N. M., Frey, M. and Iraldo, F. 2013. EMAS and ISO 14001, the differences in effectively improving environmental performance. Journal of Cleaner Production, 68, 165–173.

Tibbs, Hardin B.C. 1992. Industrial ecology: an environmental agenda for industry. The Whole Earth Review; Winter.

Tibbs, J. 1992. Industrial ecology: an environmental agenda for industry. Whole Earth Review.

Trumpp, C. and Guenther, T. 2017. Too little or too much? Exploring U-shaped relationships between corporate environmental performance and corporate financial performance. Business Strategy and the Environment, 26, 49–68. https://doi.org/10.1002/bse.1900.

Vallet, F., Eynard, B., Millet, D., Mahut, S. G., Tyl, B. and Bertoluci, G. 2013. Using eco-design tools: An overview of experts' practices. Design Studies, 34(3), 345–377. https://doi.org/10.1016/j.destud.2012.10.001.

Wagner, M. 2005. How to reconcile environmental and economic performance to improve corporate sustainability: corporate environmental strategies in European paper industry. Journal of Environmental Management, 76, 105–118.

Wang, N., Ma, M., Wu, G., Liu, Y., Gong, Z. and Chen, X. 2019. Conflicts concerning construction projects under the challenge of cleaner production–case study on government funded projects. Journal of Cleaner Production, 225, 664–674. https://doi.org/10.1016/j.jclepro.2019.03.315.

Yuan, K. -H. and Bentler, P. M. 2005. Asymptotic robustness of the normal theory likelihood ratio statistic for two-level covariance structure models. Journal of Multivariate Analysis, 94, 328–343.

Yuksel, M. 2008. An empirical evaluation of clean production practices in Turkey. Journal of Cleaner Production, 16, 50–57.

Zeng, S. X., Meng, X. H., Yin, H. T., Tam, C. M. and Sun, L. 2010. Impact of cleaner production on business performance. Journal of Cleaner Production, 18, 975–983. https://doi.org/10.1016/j.jclepro.2010.02.019.

Zhang, A., Moffat, K., Lacey, J., Wang, J., González, R., Uribe, K., Cui, L. and Dai, Y. 2015. Understanding the social licence to operate of mining at the national scale: A comparative study of Australia, China and Chile. Journal of Cleaner Production, 108, 1063–1072. https://doi.org/10.1016/j.jclepro.2015.07.097.

Zobel, T., Groß, B., Fieg, G. and Wonzny, G. 2005. Ganzheitliche Optimierung eines industriellen Prozesses mit evolutionaren Algorithmen. Chemie Ingenieur Technik, 77(7), 932–7.

Zobel, T., Groß, B., Fieg, G. and Wonzny, G. 2006. Ganzheitliche Optimierung eines industriellen Batch-Prozesses mit evolutionaren Algorithmen. Chemie Ingenieur Technik, 78(9), 1319–21.

CHAPTER 8

Partnership for Sustainable Development

Experiences From Manizales, Colombia, to Improve Environmental Education Initiatives

Paola Andrea Calderón-Cuartas,[1,*] *Javier Mauricio Naranjo-Vasco,*[2] *Cristian Moreira-Segura*[3] and *Wilmar Osorio-Viana*[4]

Introduction

The environmental crisis in the world is a reality. Climate change, air, water and soil pollution, destruction of natural areas, loss of biodiversity and inadequate waste management have affected most of Earth's ecosystems. Only a quarter of the planet's surface is outside the impact of anthropic action. By 2050, it will have declined to one-tenth (WWF International 2018). Natural resources are running out and socio-ecological systems suffer the impacts of 'The Great Acceleration', a phenomenon of economic development that has caused in recent decades an exponential increase in the demand for materials and energy, radically changing the planetary balance (Steffen et al. 2015).

[1] Doctorado en Ciencias Naturales para el Desarrollo (DOCINADE), Instituto Tecnológico de Costa Rica, Universidad Nacional de Costa Rica, Universidad Estatal a Distancia, Costa Rica. Universidad Católica de Manizales, Carrera 23 No. 60-63 Manizales, Colombia.

[2] Research Group in Technological and Environmental Developments, Universidad Católica de Manizales, Carrera 23 No. 60-63 Manizales, Colombia.

[3] Doctorado en Ciencias Naturales para el Desarrollo (DOCINADE), Instituto Tecnológico de Costa Rica, 18 kilómetros al norte de Ciudad Quesada, Carretera a Fortuna, Santa Clara, San Carlos, Alajuela, Costa Rica.

[4] Department of Chemistry, School of Agrifood and Forestry Science and Engineering (ETSEA), University of Lleida-Agrotecnio Center, Av. Alcalde Rovira Roure 191, 25198 Lleida, Spain.

Emails: jnaranjo@ucm.edu.co; cmoreira@tec.ac.cr; wilmar.osorio@udl.cat

* Corresponding author: pcalderon@ucm.edu.co

This environmental problem is a consequence of the extractive model of the linear economy, which has generated unsustainable production and consumption processes, evidenced among others, in the global increase in waste generation. Current levels of global urban solid waste generation are approximately 1.3 billion tons per year and are expected to increase to approximately 2.2 billion tons per year by 2025. In Latin America and the Caribbean, the total amount of waste generated by year is 160 million tons with per capita values ranging between 0.1 and 14 kg/person/day and an average of 1.1 kg/person/day (ONU Medio Ambiente 2018).

The problems of economic, social, cultural and environmental development in Latin American and Caribbean countries require research with integrated approaches that take advantage of advances in science, technology and innovation and consider the particularities of the context (Osorio Viana et al. 2019). The solution to these problems requires alliances between different institutions and productive sectors, constituting these in the bridge that allows organizations to collectively achieve the transformations that sustainable development requires (Nelson 2017).

The Colombian strategy for the implementation of the Sustainable Development Goals (SDGs) (Consejo Nacional de Política Económica y Social and Departamento Nacional de Planeación 2018) highlights the importance of alliances for the comprehensive management of knowledge, a knowledge that allows improving processes and making decisions for the fulfillment of the SDGs. Likewise, the strategy recognizes the circular economy as an alternative to advance toward sustainable development.

The transition to a circular economy that allows progress toward sustainable development requires several lines of work. Firstly, a political and legislative framework that structures and guides the actions of the productive sector; secondly, a commitment on the part of the different actors, who in their different roles, must be involved in the processes of production and sustainable consumption.

Achieving society's commitment to the challenges posed by the circular economy requires processes of environmental education and management. An integrative approach is needed that disseminates the principles and values of each context while promoting environmentally responsible behavior (Calderón Cuartas et al. 2019).

In Colombia, the Ministry of Environment and Sustainable Development has established a solid normative policy framework, which favors the transition to a circular economy and low carbon development including, among others, the following agendas:

1. National Policy for Environmental Education (Ministerio de Educación Nacional and Ministerio de Educación Nacional 2002).
2. National Policy for Sustainable Production and Consumption (Ministerio de Ambiente 2011).
3. National Policy for Integrated Solid Waste Management (Consejo Nacional de Política Económica y Social and Departamento Nacional de Planeación 2016).
4. The National Climate Change Policy (Ministerio de Ambiente y Desarrollo Sostenible 2017).
5. The National Circular Economy Strategy (Saer and Romano 2019).

These policies aim to strengthen the articulation between production and consumption processes to increase the circularity of materials and energy, reduce the generation of solid waste, increase recycling levels and reduce the pressure of the country's landfills (Consejo Nacional de Política Económica y Social and Departamento Nacional de Planeación 2016, SISPD and DNP 2017).

Participation and collaborative action among stakeholders are an essential factor in advancing any environmental management process for sustainable development (Nelson 2017, UN 2015). These processes must be worked from different levels such as citizenship, social organizations, government entities, companies and educational institutions among others, allowing different actors to make decisions oriented toward sustainability from each of their roles. A key aspect in these processes is raising awareness for changes in individual and collective behaviors, which to be effective must be done through environmental education programs that in addition to information and knowledge about the topic of interest include the competencies for action (Adomßent et al. 2014).

In Manizales, a city in the Colombian coffee region, each year there is an increase in per capita generation of solid waste, while recycling is scarce. These events constitute important causes of environmental deterioration and the main obstacles to sustainable development (Programa Manizales Cómo Vamos 2018). For 2018, the per capita generation of waste in the city was 1.03 kg per day, 69% more than what was produced in 2008 (Programa Manizales Cómo Vamos 2019). Currently, La Esmeralda landfill receives 540 tons per day of urban solid waste from Manizales and other municipalities. Only 1.4% of the city's waste is recycled, while Colombia's goal in SDG 12 is to achieve a national rate of waste recycling, i.e., 18% by 2030. (Consejo Nacional de Política Económica y Social and Departamento Nacional de Planeación 2018).

This city is characterized as a university city that has the SUMA Alliance (Manizales University System) institutionalized in 2010 and is formed by six universities with the purpose of consolidating an academic and administrative organization that strengthens each of its missionary functions as well as reinforcing education on a complete perspective of local and global realities (Manizales University Campus 2015).

In the context of Manizales, Universities are called to lead alliances that address local environmental problems, including those related to Solid Waste Management (SWM). Addressing these complex problems from collaborative action has several specific goals:

a. The training of professionals with a high degree of environmental responsibility with society.
b. The conformation of sustainable campuses, which quickly experience an environmental ethic and the progressive establishment of an environmental culture inside and outside the University.
c. To drive processes of transformation and sustainable development in the territories.

Facing local environmental problems requires Environmental Education (EE) processes that involve effective elements that lead communities to have concrete practices and actions, aimed at the use and conservation of the environment. Alliances for sustainable development must bet on an EE that allows actors to solve environmental problems in their context and articulate individual, inter-institutional and intersectoral responsibilities.

In this chapter, we will present, in the first place, a theoretical perspective that explores the importance of interinstitutional and intersectoral alliances in the university context as well as the urgency of moving toward Effective Environmental Education (EEE) processes, which arise from the identification of environmental problems in the context and that favor the participation of the involved stakeholders. Subsequently, we will present three experiences of networking to promote the effectiveness of environmental education processes in the city of Manizales. First, the case of the interuniversity alliance for sustainable development: SUMA Ambiental (for its acronym in Spanish). This alliance has allowed progress in a research process that aims to consolidate an environmental education plan for the integrative solid waste management in Higher Education Institutions (HEI) that are members of the alliance. Another experience is the case of intersectoral alliances between a university, a cleaning company and an environmental foundation that has concluded in the creation of (i) 'Chair in Production and Sustainable Consumption: From Values to Actions' and (ii) 'Waste Separation Program in Educational Institutions of Manizales'. Finally, the case of alliances between different organizations to form a public policy strategy: The Inter-institutional Committee of Environmental Education of Manizales (CIDEAMA for its acronym in Spanish).

Literature Review

The need for partnerships for sustainability in the university context

Historically, global organizations such as the United Nations have promulgated the need for sustainability in higher education (Unesco-Pnuma 1978); there have been multiple international declarations in which universities are committed to addressing environmental issues to contribute to sustainable development policies and be an example for society from an institutional environmental performance, including the Intergovernmental Conference on Environmental Education (UNESCO-Pnuma 1978), the Declaration of Talloires made by the Association of University Leaders for Sustainable Development (ULSF 1990) and the Regional Conference on Higher Education in Latin America and the Caribbean (IESALC-UNESCO 2008).

There is also a generalized call for inter-university cooperation with the purpose of generating sustainability networks that strengthen the environmental commitment of higher education (Vallejo 2013, Campos and Camacho 2015, Unesco-Pnuma 1978). Evidence of this is the increasing participation of Colombian universities in national, Latin American and international networks for sustainability, including the Colombian Network of Environmental Training, the Environmental Network of Sustainable Universities, the University Union in Sustainable Production and Consumption, Alliance of Ibero-American Networks of Universities for Sustainability and the Environment (Sáenz and Benayas 2015). Likewise, the annual increase of Colombian universities participating in the Greenmetric World University Ranking demonstrates the important dynamics that the country has in this area. In 2019, 45 Colombian universities participated in this ranking, 309% more than in 2015 (UI Greenmetric 2019).

In university contexts, networking with partners from other educational institutions and organizations from different sectors becomes a key factor in ensuring the adequate achievement of the key functions of higher education, teaching, research and outreach/public service.

The main purpose of these alliances is to contextualize the academy to close the gap between theoretical knowledge and the realities that professionals must face (Irazábal et al. 2015). Among the success factors in forming alliances and networking in different types of institutions, the following have been identified (Jones et al. 2016, Nelson 2017).

1. The establishment of common objectives and specific commitments by each institution.
2. The generation of an agenda for action, providing resources and guaranteeing the competencies of team members.
3. The revitalization of leadership, communication and measurement mechanisms to assess progress.

The alliance between educational institutions is common in the academic context. A model proposed for collaborative work between Basic-Media-Higher Education Institutes to achieve an effective education of teachers (Jones et al. 2016) identifies the collaboration, coordination and communication as facilitators of the growth process. These facilitators (foster with the behavior, expertise, attitudes and values of the people involved in the work teams) determine key aspects to achieve the purposes of the alliance such as identity, trust, praxis and good relationships. This model, with some adaptations, can be useful for different types of alliances among educational institutions and between these with other types of organizations (Figure 1). It is important to highlight people within organizations seeking collaborative work because they recognize that it increases the probability to achieve the proposed goals (Hobbs and Campbell 2018).

Environmental Education and the Factors that Limit its Effectiveness

In recent years, environmental concern among citizens has increased in general due to environmental information disseminated by the media and by educational institutions; but paradoxically, unsustainable production, consumption and post-consumption practices have also increased. For example, it has not been possible to establish recycling as a cultural practice (Páramo 2017).

For four decades, the global need for EE to adopt a holistic approach to address and solve environmental problems from ecological, social and cultural aspects has been identified (Unesco-Pnuma 1978). However, the 2030 agenda for sustainable development shows that this challenge persists. It is necessary to ensure that citizens have the knowledge, skills, values and attitudes to make informed decisions and to play an active role in facing and solving local and planetary problems (United Nations 2015).

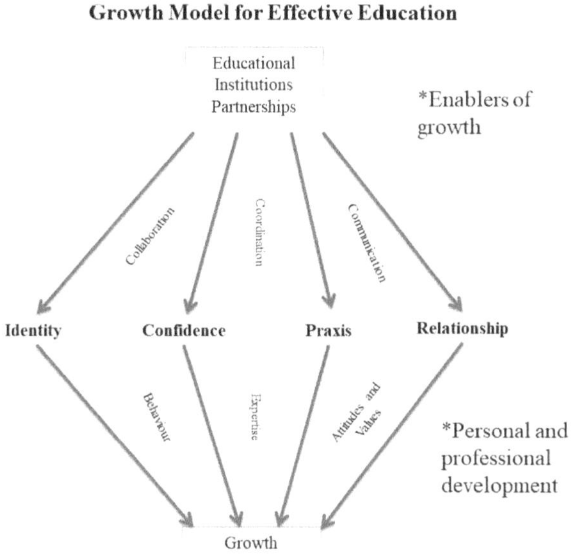

Growth Model for Effective Education

Educational Institutions Partnerships

*Enablers of growth

Collaboration
Coordination
Communication

Identity
Confidence
Praxis
Relationship

Behaviour
Expertise
Attitudes and Values

*Personal and professional development

Growth

Figure 1. Growth model for effective education. Source: Adapted from Jones et al. 2016. Successful university-school partnerships: An interpretive framework to inform partnership practice. Teaching and Teacher Education, 60, 108–120.

One of the challenges that sustainable development poses to higher education is to contribute to training and environmental education in university settings, through formal and non-formal processes (Mora 2009). Environmental education in any of its modalities aims to contribute to the acquisition of awareness, values and attitudes, techniques and environmental behaviors and to adequately address environmental issues and contribute to sustainable development (Rivas 2014). And taking into account that each sector of society contributes differently to the solution of environmental problems, it is necessary to adopt various strategies and methodologies for working with different groups (Coya 2001).

Findings of environmental psychology help explain why we cannot always perform actions, despite our good intentions (UNEP 2017). The gap between intentions and actions is the same gap that has limited effective environmental education. Among the limiting factors of Environmentally Responsible Behaviors (ERB) are the conceptual, attitudinal and control barriers that become obstacles and inertia that impede the effectiveness of environmental education proposals (Mora 2009).

The conceptual barriers are those that originate from the teachers' misconceptions about the objectives of the EE (Ham and Sewing 1988). Attitudinal barriers are those that are formed from the erroneous or biased perceptions of the teachers themselves, which have repercussions on the sensitivity and attitude of the students (Ham and Sewing 1988). On the other hand, the control barriers are those originating from the perception of the facilitator about the lack of resources and opportunities to achieve ERB (Charpentier 2004).

The fact that the EA traditionally uses methodologies of a purely informative-communicative nature is evidence of the conceptual barriers that arise. These methodologies can generate a cognitive effect on people but not necessarily generate changes in attitudes and behavior (Pauw et al. 2015). To promote pro-environmental actions, such as those related to recycling, it is necessary to act on the attitudes of the different stakeholders involved in the educational process (Melero et al. 2013). The development of pedagogical processes to provoke the transformation of individual and collective behaviors must also include specific competencies for informed decision making (Adomßent et al. 2014).

Likewise, it is necessary to transcend toward the measurement of the impact to know the effectiveness of the EE processes, identifying and facing the obstacles that prevent achieving their purposes. In this sense, educational programs that are designed in universities to promote ERB, Pro-Environmental Behaviors (PEB) or PEA must necessarily be formulated based on the diagnosis of the competencies required by the different participants. All these, according to the local characteristics of the context, the type of environmental problem addressed, the possibilities of action according to the role and the barriers identified.

It is important to understand that environmental education processes are highly complex and that to be effective, they require transcending the simple vision of expecting immediate behavioral changes from educational interventions. Interdisciplinary work and articulation with scientific education are required to achieve the purposes of environmental education. It is necessary for environmental educators to analyze how people develop competencies to holistically understand environmental problems and contribute to solving them (Dillon et al. 2018).

Models for Effective Environmental Education

To achieve an EEE process, it is suggested the application of models to:

1. Relate awareness, knowledge and environmental attitudes with context-specific variables, and in this way recognize the specific conditions that favor and limit responsible environmental behaviors (Oztekin et al. 2017, Palacios et al. 2015, Thomas and Sharp 2013).
2. Identify the competencies necessary for the solution of problems related to the use and management of resources (Unesco-Pnuma 1978) and provoke the transformation of individual and collective behaviors from informed decision making (Adomßent et al. 2014).

Likewise, it is important that these processes are validated and evaluated, and in this way, they propose solutions that allow closing the gaps that exist between environmental beliefs and behaviors (Lopez and Guerrero 2017, Páramo 2017).

In this chapter, we present nine models that PEB has studied from psychology and EE, including waste minimization and recycling. Each of these models was characterized, identifying its variables and establishing whether it analyzes two variables that can be determining factors in the effectiveness of EE: the context and the competencies.

Table 1 presents, in chronological order, the models or theories applicable to environmental education for the integrated management of solid waste. This exercise determined the selection of the model to apply for Case 1: Integrated Solid Waste

Table 1. Models or theories applicable to environmental education for behaviors related to ISWM.

Model/theory applicable to EA for GIRS	**Description**	**Variables that relate and contrast with established criteria: 1. Context 2. Competencies**
1. Norm Activation Model (Schwartz 1977)	The basic premise of this model is that moral or personal norms are direct determinants of pro-social behavior, which in turn motivates PEB. Several studies provide evidence that moral norms contribute to an explanation of PEB such as recycling (Guagnano et al. 1995).	The behavior depends on the personal norms that are affected by the following variables: - Awareness of consequences - Need conscience - Situational responsibility - Effectiveness - Skill - Denial of responsibility **1. Does not consider context variables.** **2. The skill mentioned can be considered as competencies.**
2. The Theory of Planned Behavior (Ajzen 1991)	Intentions to perform behaviors of different types can be predicted with great precision from attitudes, subjective norms and perceived control of behavior. This theory is a complement to the Theory of Reasoned Action, which studied at the time, the influence of the social group on individual pro-environmental behavior. It is one of the most used theories from social psychology to analyze behaviors for recycling (Lizin et al. 2017, Oztekin et al. 2017, Park and Ha 2014, Tonglet et al. 2004).	The execution or non-execution of a behavior is the result of the intention. The theory relates to the variables: - Attitudes - Subject rules - Perceived behavior control These variables in turn are the product of a group of beliefs in their order: - Behavioral beliefs - Regulatory beliefs - Control beliefs **1. Does not consider context variables.** **2. Does not consider competence variables.**
3. Responsible Environmental Behavior Model (Bamberg and Möser 2007, Hines et al. 1987)	Responsible or pro-environmental environmental behavior is a mixture of self-interest and prosocial reasons: concern for other people, the next generation, other species or complete ecosystems. Studies based on this model have ratified average correlations between psychosocial variables and pro-environmental behavior, confirming that attitude, behavioral control and moral norm are the main predictors of behavioral intention for recycling (Vicente et al. 2013).	The following variables were associated with responsible environmental behavior: - Knowledge of problems - Knowledge of action strategies - Locus control, - Abilities, - Verbal commitment - Sense of responsibility of an individual. **1. Does not consider context variables.** **2. The variable 'knowledge of action strategies' can be considered as competence.**

Table 1 Contd. ...

...Table 1 Contd.

Model/theory applicable to EA for GIRS	Description	Variables that relate and contrast with established criteria: 1. Context 2. Competencies
4. Focus Theory of Normative Conduct (Cialdini et al. 1990)	The theory focuses on norms, which systematically affect human action, descriptive norms and prescriptive norms. The former refers to what people normally do and motivate the behavior by providing information about what is more normal or typical; the latter refers to what is normally approved or disapproved within a culture and are motivated by the promise of rewards or punishments. In contrast to the descriptive norms, which specify what is done, the prescriptive norms specify what must be done (Alonso 2012).	The theory states that the frequency with which a transgression occurs increases the occurrence of it. To demonstrate this statement, the authors conducted three studies in which they verified that the descriptive norms have an important effect on people since more papers are thrown in a place on whose floor there are already papers than in a place where the floor is clean. Disruptive behavior increases if someone is seen throwing a paper to a floor that is already dirty and decreases when the floor to which the paper is thrown is clean. **1. Does not consider context variables. 2. Does not consider competence variables.**
5. Positive Environmental Action Model (Emmons 1995)	A Positive Environmental Action (PEA) is a strategy that involves a decision, planning, implementation and reflection process by an individual or group. It differs from the environmental behavior responsible for empowerment and self-determination. The PEA depends on both internal and external motivating factors, which could prevent a person from executing a behavior, for example, separating solid waste (Campos and Camacho 2015).	Variables that relates: - Environmental concepts - Action skills and procedures - Attitudes and sensitivity - Empowerment and ownership - Positive environmental action **1. Does not consider context variables. 2. The variable 'Action skills and procedures' can be considered as competence.**
6. Sustainability Literacy (Murray and Cotgrave 2007, Parkin et al. 2004, Stibbe 2009)	Literacy for Sustainability is articulated in Education for Sustainable Development. This type of education prioritizes knowledge to encourage students to develop critical, holistic, systemic and interdisciplinary thinking skills. One way to evaluate this type of literacy is the Sulitest (Sustainability Literacy Test[1]), created in 2017 to evaluate the conceptions of sustainability of university students and involve higher education institutions in monitoring the impact of the integration of sustainability into its pedagogy (Décamps et al. 2017).	Sulitest is an articulated tool for sustainable development goals. The questionnaire includes questions about: -Knowledge that seeks to identify the conceptions of sustainability of students on campus. Although it is planned to measure skills and mentalities, it is still in the process of expanding the questionnaire. **1. Does not consider context variables. 2. Competence variables are considered as 'skills'; however, the instrument is currently focused on knowledge.**

[1] To date, more than 120,000 tests performed by university students from more than 840 institutions of higher education are reported. As Sulitest is applied by various cohorts of apprentices in the world, it is useful in generating data that allows understanding behaviors. https://www.sulitest.org.

Table 1 Contd. ...

...Table 1 Contd.

Model/theory applicable to EA for GIRS	**Description**	**Variables that relate and contrast with established criteria: 1. Context 2. Competencies**
7. Framework for Assessing Environmental Literacy (Hollweg et al. 2011).	The term Environmental Literacy (EL) emerged in 1968 in the United States and has evolved into a process of developing environmental competencies and evaluating environmental education processes (Roth 1992). The EL evaluation model states that a person with environmental knowledge and emotional dispositions can make responsible decisions for the resolution of environmental problems.	The model relates the following variables: - Context - Environmental knowledge - Provisions - Competencies - Responsible environmental behavior **1. Consider context variables.** **2. Consider competence variables.**
8. Sustainable Behavior Model (Juárez Nájera 2015)	A new psychological model framed in the approach of Education for Sustainability. It is an adaptation of the Activation Model (Schwartz 1977) including elements of other models such as the Standard-Belief-Values. The model differs from others because it is based on a complete range of universal values. The model aims to identify the variables that best explain the disposition to sustainable behavior as well as the most relevant perceptions that can influence behavior change.	The model relates the variables of sustainable behavior: - Universal values - Knowledge of the consequences - Attribution of responsibilities - Personal intelligence 1. Validation questionnaires include questions specific to the context. 2. It does not include competence variables directly, but personal intelligence is related to skills. The model is tested by the author from an exploratory-confirmatory study in a university context; however, no other studies have been reported that have applied it.
9. The Identity-Based Environmental Education Model (Mcguire 2015)	The author proposes the model as a possible new approach to address environmental problems that are increasingly global in their impacts and diffuse in their causes. All environmental problems arise as a function of consumption and the identity-based environmental education model presents an approach that stands out to address this fundamental relationship. Identity-based environmental education allows the ecological rectitude of behavior to be determined by affection and emotion, rather than an abstract and complicated projection of how a particular action will affect natural resources.	Unlike the models/theories or approaches outlined above, which have studies that have been empirically validated through multiple investigations, this model constitutes a proposal with a solid theoretical basis, which has yet to prove its validity.

Source: Prepared by the authors

Management – ISWM of the SUMA Alliance. It is important to mention that most of these models have correlational studies based on surveys or questionnaires where they establish which variables are related to environmentally responsible behavior.

In Latin America, there are few studies that have applied these models and theories in local contexts with practical implications to support effective environmental education strategies. Among the published research, we can cite the authors Campos and Camacho (2015), Juárez Nájera (2015) and Moreira et al. (2015).

As evidenced, the EL model is the model that meets the two established criteria. The EL model applies to both formal and non-formal environmental education and refers essentially to the development of competent citizenship to understand the functioning of natural systems, the interaction between nature and humans and the ability to act in critical environmental issues (Roth 1992). Among the conclusions and recommendations made by different studies that have evaluated EL, the following stand out:

a. The "knowledge" has a weak relationship and a minor contribution to explaining the variation in the environmental behavior of students. A much stronger relationship was found between values and attitudes and behaviors (Arnon et al. 2015).
b. The competence-based approach used by the EL model is useful to identify the capacities required in individuals to act in conditions in which a problem must be resolved (Corral-Verdugo 1994). The competent performance of effective action implies the mobilization of knowledge, cognitive and practical skills as well as social and behavioral components, such as attitudes, emotions, values and motivations. Competence is a holistic notion; therefore, it is not reducible to its cognitive dimension, making it clear that the terms "competence" and "ability" are not synonymous (Hollweg et al. 2011)
c. It is necessary that the educational perspective on the issue of waste management is based on a preventive approach to avoid that promotion of recycling involves practices that increase consumption since students do not connect their consumption with environmental impacts (Goldman et al. 2013).

Experiences of Alliances in Manizales

Interuniversity Alliance for Sustainable Development: SUMA Ambiental

Suma Ambiental emerged in 2017 as a line of work of the SUMA Alliance. This new interaction framework brings together the leaders and environmental managers of the six universities that make up the alliance: Catholic University of Manizales (UCM for its acronym in Spanish), University of Manizales, Autonomous University of Manizales, Luis Amigó Catholic University, University of Caldas, National University of Colombia, Manizales Campus with the purpose of joining efforts in favor of the environmental management of the university campuses and of the environmental educational processes that Manizales requires to consolidate itself as a Learning City and Sustainable University Campus.

This alliance has two lines of work: University Environmental Management and Academic Board, which support the missionary functions of Universities, and

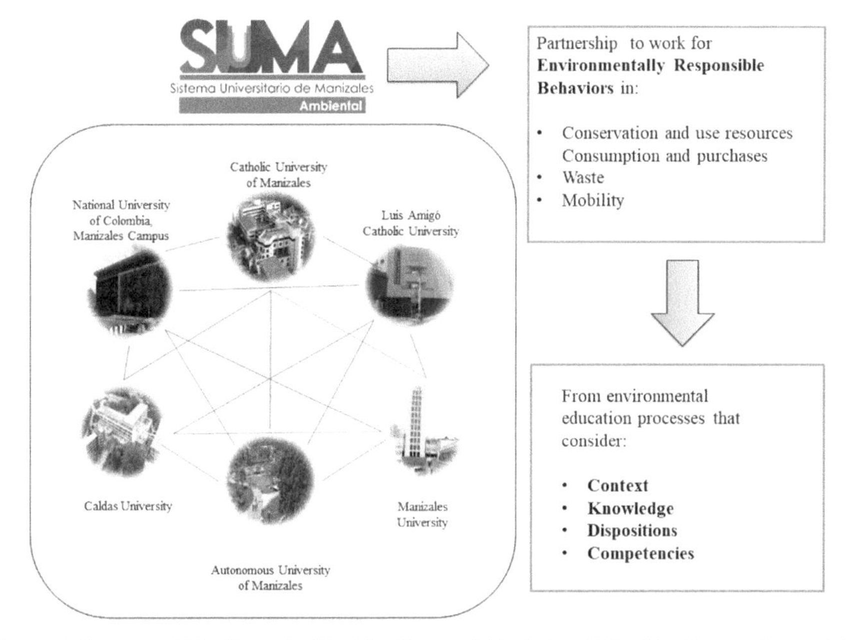

Figure 2. Purposes of the Suma Ambiental Alliance of Manizales, Colombia. Source: Prepared by the authors.

Teaching, Research and Social Outreach (Figure 2). As the first research exercise of the Alliance, the application of a diagnostic instrument that identified the environmental dimension has been incorporated into each of the HEI as a basis for understanding environmental management and education processes, especially those related to solid waste management, which is one of the biggest challenges to achieve sustainable campuses (Ghazvinei et al. 2017).

In methodological terms, the adapted instrument of Sáenz et al. (2017) obeys to the descriptive and contextual phase of the project, which was validated with expert criteria by environmental leaders of the Costa Rican Network of Sustainable Educational Institutions (REDIES for its acronym in Spanish).

The questionnaire is structured in five areas according to what was proposed by Román (2015) for the University Environmental Systems (UES) and what was established by Sáenz et al. (2017) and Callejas et al. (2018) on Institutional Environmental Systems (IES). The IES is conceived as an integrated management model that allows consolidating and organizing the guidelines, norms, orientations, plans, programs, projects, activities, goals, indicators, resources and managers that guide and regulate the planning and management of a HEI toward sustainability.

Holguín (2017) stated that the UES includes the formation of environmental culture in the university community from an institutional and territorial context, favoring environmental management from a systemic approach. On the other hand, Sáenz et al. (2017) and Callejas et al. (2018) expanded the concept of the UES, calling it the IES for the purpose of evaluating the institutionalization of the university environment commitment based on the components or areas presented in Figure 3.

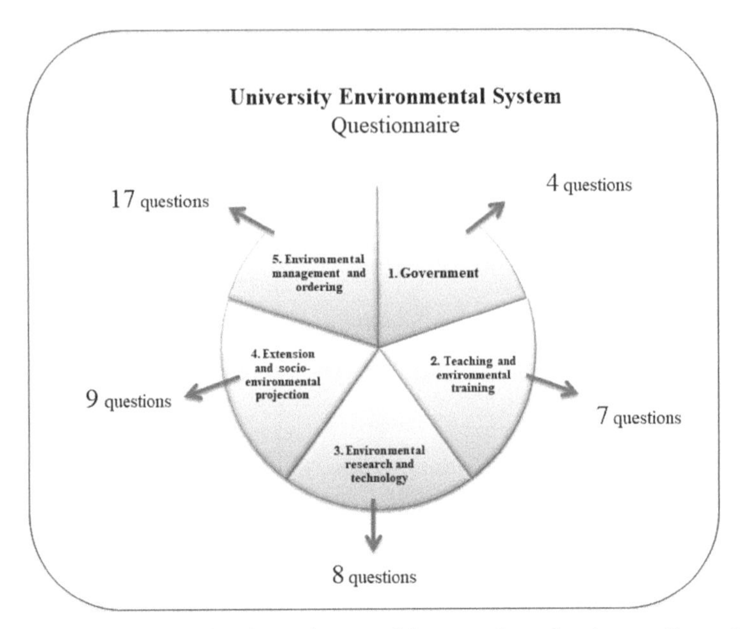

Figure 3. Dimensions of the University Environmental System and questions by area. Source: Prepared by the authors.

The instrument, consisting of a total of 45 base questions,[2] was answered by five universities of the six SUMA universities. In addition to obtaining information for this project, the instrument is expected to allow the participating institutions to systematize the different strategies applied as well as identify strengths and weaknesses as a tool for decision-making and formulation of improvement plans in environmental processes.

The main results are presented below:

1. Government

   All SUMA universities have a high level of institutionalization of environmental commitment. This is due to each university has an institutionalized environmental policy, which declares itself to society its decision to face environmental challenges. Most universities have created positions to manage environmental issues from a technical or administrative office led by environmental managers.

2. Teaching and environmental training

   In all universities, the incorporation of the environmental perspective is reported mainly through curricular subjects.

   Most universities offer at least one specific undergraduate or graduate curricular program in the environmental component.

[2] Mainly closed questions type checklist to know if the HEI has the strategy investigated or not. Likewise, complementary questions were raised to broaden the answer to each base question, allowing to characterize each scope of action of the university environmental systems.

It was identified that not all SUMA universities have training strategies for teachers and students to strengthen the environmental perspective in teaching and curricula. There is little articulation between UES and environmental training processes.

3. Environmental research and technology

   It is reported that all universities have at least one research group that manages environmental knowledge. There exists a low degree of articulation and transfer of the results in environmental research to the development of university campuses.

4. Socio-environmental outreach

   The SUMA universities have a high level of socio-environmental responsibility because they all report various strategies, such as environmental projects in communities, University-Company-State relationship, active participation in academic networks and non-formal environmental education.

5. Environmental management and ordering

   Among the environmental processes of least implementation in universities, it is possible to mention:

   - The acquisition of goods and services with sustainability criteria and sustainable mobility.
   - Although all universities report having implemented some plan, program or action for the ISWM, environmental managers show difficulties in the adequate separation at the source, showing weaknesses in the EEE. Not all universities have indicators of generation and use of solid waste.

Figure 4 presents the results corresponding to 40 questions. Five open questions that correspond to environmental management and the ordering of the campus are excluded.

Intersectoral Alliance for Education in Production and Sustainable Consumption

Contributing to the management of knowledge in production, sustainable consumption and the integral management of solid waste is a challenge that has been embraced by the UCM. All this, considering the purposes of the Environmental Engineering Department and the Research Group on Technological Developments and Environment (GIDTA for its acronym in Spanish) as well as the social responsibility inherent in its model of Social Management and promulgated by the University Environmental Policy based on the Encyclical Letter Laudato Sí on the care of our common home.

The Environmental Thought Center Foundation (CEPA for its acronym in Spanish) was created in 2014 by the Metropolitan Cleanliness Company (EMAS for its acronym in Spanish) with the mission of contributing to the training of environmentally responsible citizens through environmental research and education (Calderón-Cuartas et al. 2019).

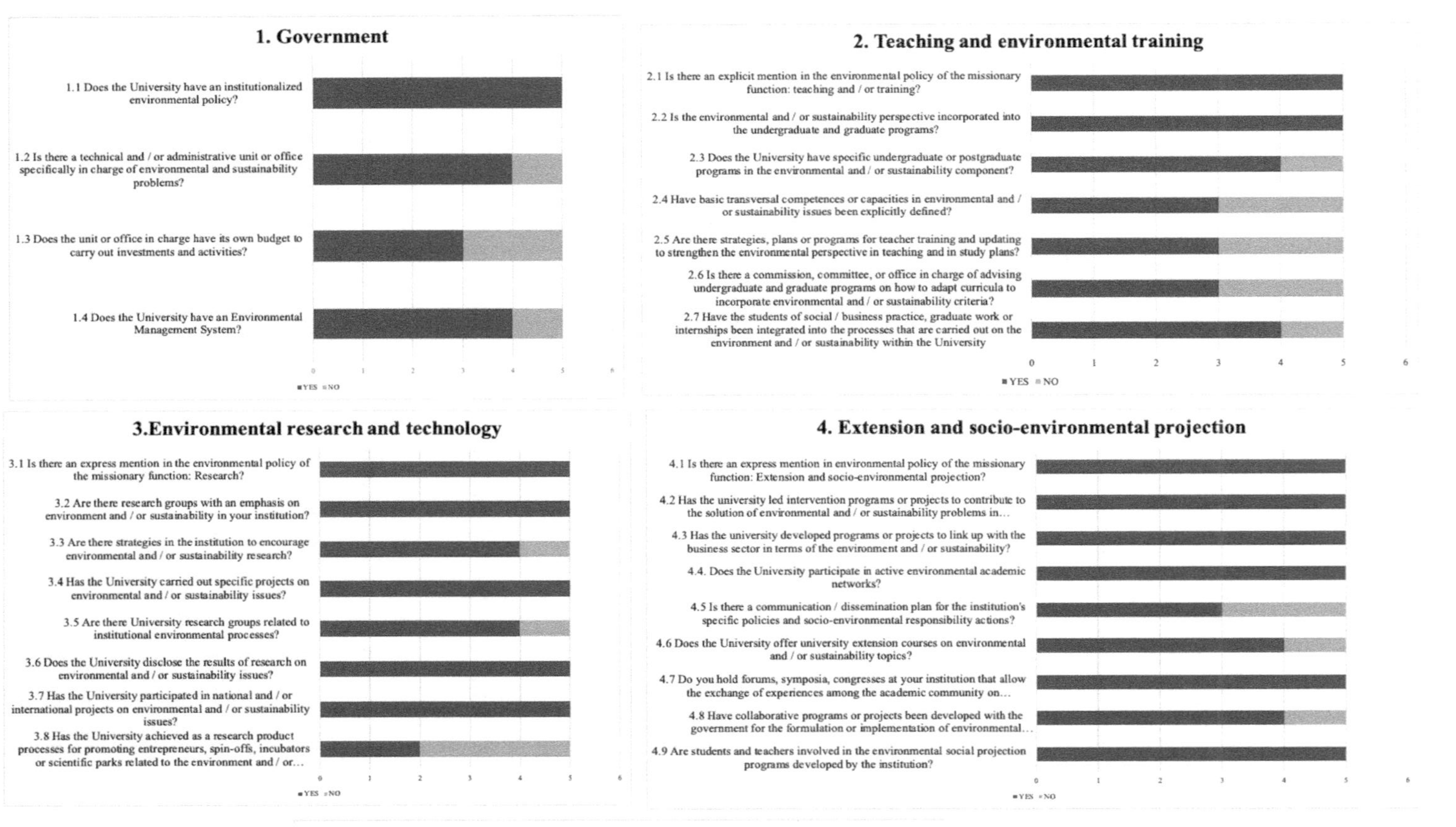

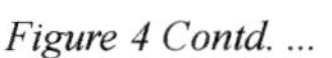
Figure 4 Contd. ...

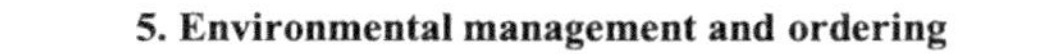

...Figure 4 Contd.

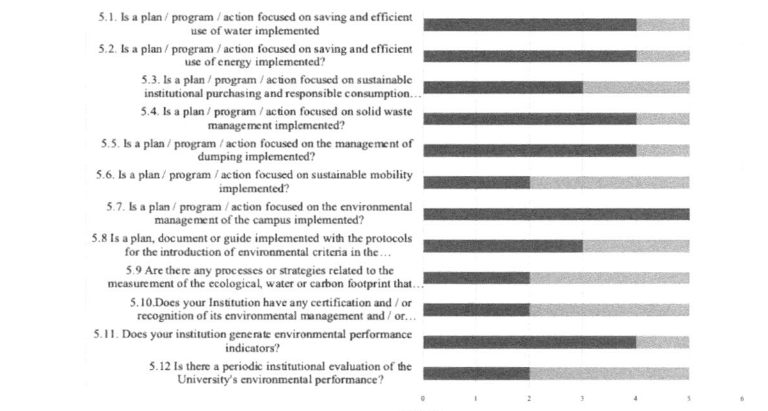

Figure 4. Questionnaire results that identified how the environmental dimension has been incorporated in SUMA universities. Source: Prepared by the authors.

With the purpose of joining efforts to achieve these common purposes, in 2015 the alliance between the UCM, CEPA and EMAS emerged. Since this alliance, two non-formal environmental education strategies have been consolidated: The Chair in Sustainable Production and Consumption and the Waste Separation Program in Educational Institutions (SERES for its acronym in Spanish).

With the motto 'From values to actions', the Chair in Sustainable Production and Consumption is based on the National Policy of the same name and in SDG 12. Moving from values to actions means closing the gap between intentions of action and behaviors since there is often a disconnection between them due to different barriers (UNEP 2017). It is a call to collectively identify the barriers that limit the transformation of thoughts, beliefs and knowledge into behaviors and actions for sustainability (Calderón-Cuartas et al. 2019).

From the methodological point of view, both the Chair and the SERES Program have been based on the principles of the Circular Economy and the Life Cycle Approach (LCA), which considers the environmental aspects and impacts of the goods and services in all phases (material and energy extraction, production, use, distribution, final disposal, reuse and recycling). The LCA implies an essential purpose that focuses on awareness, responsibility and commitment to work individually and collectively in the prevention and resolution of environmental problems that arise from the processes of transformation of raw materials into products and services and its subsequent disposal (Calderón-Cuartas et al. 2019). In this sense, the conception of the Circular Economy and the LCA are tools that allow different stakeholders to understand the environmental impacts that are generated around the products and services consumed daily. These tools establish new criteria that are evidenced in environmentally responsible behaviors related to the decrease in consumerism, institutional purchases with sustainability criteria, separation of solid waste at the source and recycling among others.

The chair is done through various modalities: workshops, forums, expert conferences, guided tours and waste management fair among others. After three years (2016–2019), more than 15 meetings have been held within the framework of the Chair, calling more than 1,000 participants among members of the academic community, social and environmental managers and national, regional and local leaders and more than 40 speakers from universities, public and private sector institutions, non-governmental organizations, members of community action boards, 'citizen watcher', civil society groups, industry and other institutions that advance environmental sustainability initiatives.

The topics addressed in the chair have been reflected in the first book that compiles reflections and experiences on sustainable production and consumption processes, topics of great importance for the sustainable development of the region. The reflections set forth in the book called *From values to actions, a chair to promote sustainable production and consumption* and also include papers endorsed by deliberations, learning and meetings promoted since this meeting.

SERES was conceived in 2016 with the purpose of promoting the culture of responsible consumption, source separation and solid waste recycling, contributing to the environmental quality of Manizales from an educational strategy based on the life cycle approach. The program has four components, such as educational, social, logistic and economic, that aim to articulate the technical dimension of solid waste management with a socio-cultural dimension based on education and environmental culture.

The first action carried out by the program was the diagnosis of solid waste management in 60 educational institutions in Manizales through which the needs could be identified not only in environmental education but also in infrastructure and waste collection centers. This is how it was determined that in most cases, the barriers of adequate separation at the source and integral management of waste in these institutions are not the sociocultural component but a technical component. This barrier was related to infrastructure conditions of educational institutions and articulation with managers for the selective and timely collection of usable waste.

Currently, the program is in the validation phase of the didactic materials prepared from which it is expected to transcend the implementation of the educational process and measure the impact achieved through indicators that contemplate the different variables involved in the educational processes: environmental knowledge, dispositions, skills and, finally, behaviors.

Inter-governmental governmental Alliance for Environmental Education of Manizales

The National Policy of Environmental Education of Colombia established among the criteria of the EE, interdisciplinarity, interinstitutionality and intersectorality since no organization, by itself, can face an environmental problem in an integral way (Ministerio de Educación Nacional and Ministerio de Educación Nacional 2002).

Environmental education as a means to train new citizens through the establishment of environmentally responsible behaviors requires strategies for participation and cooperation at the local level. Thus, the policy created the Inter-institutional Committees of Municipal Environmental Education (CIDEAM for its acronym in Spanish) is considered as one of the main strategies to contextualize the Environmental Policy at the territorial level and properly manage the environmental education.

These committees must be led by the Municipal Education Secretary in coordination with the environmental authorities and in this case, the Regional Autonomous Corporations (CAR for its acronym in Spanish) and the Municipal Environment Secretary. Participants are organizations that have an interest and responsibility in environmental education, such as home-based public service providers, educational institutions (schools, colleges and universities), non-governmental and civil society organizations and companies among others.

In Manizales, this committee is called the Inter-institutional Committee of Environmental Education of Manizales (CIDEAMA). Figure 5 presents the organizations that make up CIDEAMA.

CIDEAMA has action fronts focused on different forms of environmental education (formal and non-formal). It aimed at various audiences: students (primary, basic, secondary, middle and higher), entities and citizens in general. The importance of this committee is that through participatory planning that responds to the environmental needs of the territory, efforts can be focused on city processes from

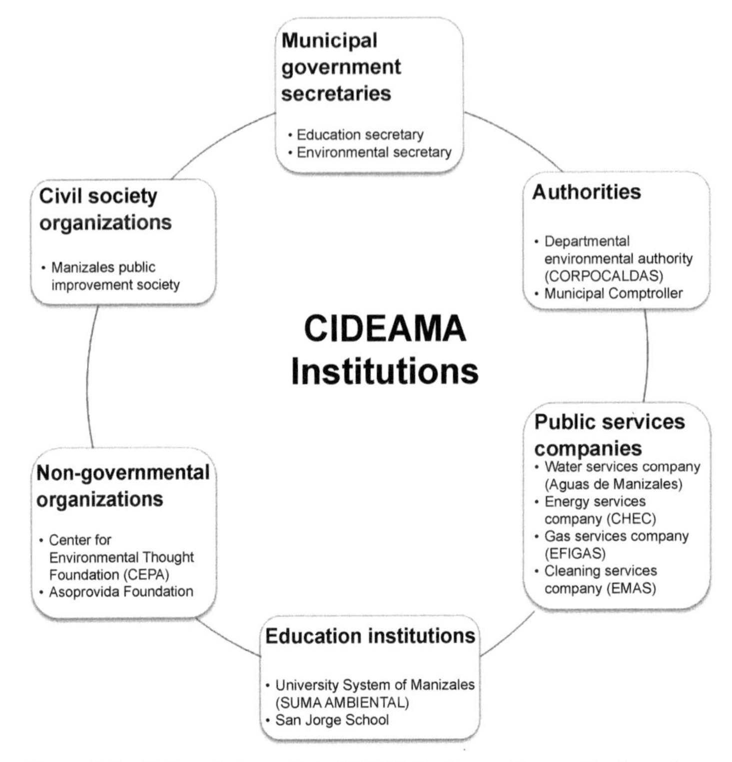

Figure 5. Institutions that constitute CIDEAMA. Source: Prepared by the authors.

a collective action that contributes to the ethical-environmental formation. Each of the institutions that participate in the CIDEAMA has particular intentionality in educational-environmental processes, whether for corporate social responsibility, community intervention, citizen commitment or for legal aspects.

The work articulated in CIDEAMA has allowed to combine efforts, resources and build a vision of Manizales as a sustainable city. In the last year, CIDEAMA has led two educational-environmental processes of great interest and impact for the community of Manizales: The Environmental Week, and the Project for the Strengthening of the Pedagogical Axis in EE of the Educational Institutions of Manizales.

The Environmental Week is constituted as a city event that takes place every year led by Corpocaldas and the Ministry of the Environment and in which the institutions of CIDEAMA joined. Comparing the impacted population between 2018 and 2019, an increase of 40% and a 30% increase in the number of activities carried out could be evidenced.

Another significant experience of CIDEAMA was a project led by the Ministry of Education and the Catholic University of Manizales with the support of the Ministry of Environment, Aguas de Manizales, Emas, Efigas, Chec and the Center for Environmental Thought-CEPA in 2019. This project was focused on strengthening the Pedagogical Axis in Environmental Education in the Educational Institutions of

Manizales. As a result, teachers involved in strategies for mainstreaming education in 56 Public Educational Institutions in the municipality of Manizales could be trained. In addition, a participatory environmental diagnosis of the territory could be carried out for the first time in the city in the 11 urban sub-areas (communes) and the rural part of the city, identifying the main problems that arise in these contexts. These workshops were attended by teachers, students, parents and community leaders who 'read' their territories and prioritized the identified environmental problems. In 63% of the communes, inadequate solid waste management was identified as the most important environmental problem in their territory. In 27% of the communes, they considered that this is a problem of medium importance. On the other hand, the participants of the rural area, identified the inadequate management of the water resource as their most important environmental problem, while the problems related to solid waste are considered as a problem of medium importance.

Conclusions

Despite the diversity of educational proposals that deal specifically with training in competencies for the protection of the environment, there is still much to do on research in the pedagogical sciences to achieve significantly more effective strategies for the education of citizens and future generations. It has been proposed in this study that an integrated approach to environmental education in which knowledge, attitudes, values and actions are combined through a critical view of the individual's context, which may be more effective in achieving a more autonomous and self-critical person concerning their daily behaviors toward the environment and their protection. But the answer is not so simple since people are not only influenced by their education but by the set of social structures that define the society in which they live. Therefore, the need arises for this social structure to be united through mechanisms that allow it to carry out coordinated and consistent actions about environmentally responsible behaviors. That is why here we have advocated promoting partnerships among organizations as a determining factor to increase the effectiveness of environmental education initiatives and the success of related sustainable development projects.

Behind this proposal, the central idea is that the individual is more susceptible to critically self-regulating its environmental behavior if: (1) he/she identifies the realities and needs of his/her daily habitat (environmental and socio-cultural context) and has developed a civic connection with his/her improvement, (2) if he/she has knowledge that allows him to understand basic aspects of the rational functioning of the natural and built world that he/she inhabits and (3) if he/she has managed to develop the ability to use that knowledge and intentions to manipulate his/her surrounding reality seeking for a collective benefit. Also, the people must be surrounded by social structures that are consistent with the goals of environmental protection and that provide tools and opportunities (legal frameworks, economic resources, infrastructure, etc.) to act in pursuit of environmental conservation in everyday life.

Taking care of the planet should be a collective responsibility and although the individual is, throughout its life, the architect of the concrete actions that contribute to protecting nature immediately or in the short-medium-or long term, it is not reasonable that the full weight of environmental protection and conservation be unloaded on the person as an individual. The educational models that promote this idea, visualizing apocalyptic futures and appealing to the individual's moral conscience to avoid them, do not know that the structure of reality is not a straight line that can be drawn from the presence of an individual to the future but rather a set of complex interactions in which individuals act dynamically and are transformed in unsuspected ways by the strong connection between natural processes in the environment, anthropogenic action on nature and the strained relationship between technology and society.

The three examples of alliances for environmental sustainability and the promotion of environmental education in the city of Manizales (Colombia) that have been presented in this study shows the suggested approach that can be applied in different communities to build that social support that will help individuals to act according to desirable principles of protection and conservation of the environment. Academia (educational institutions) must provide a strong scientific and technological foundations as well as the most suitable didactic tools for carrying out the projects; the private companies have to contribute its understanding of practical problems, its technical expertise, its technology platforms and its financial resources as well as its communications network with its clients and users of public services. Non-governmental organizations and community associations must contribute with their citizen participation mechanisms and with their power to collect and group citizen concerns and proposals. Finally, governmental organizations must guarantee the availability of democratic instruments for environmental sustainability and environmental education projects, while promoting consistency with the framework of national and international policies with which there is a social commitment, such as achieving the sustainable development goals.

Institutional alliances, as suggested by SDG 17, are essential to optimize resources, obtain greater impact and build processes of transformation of society within the framework of sustainable development.

Environmental education, being a complex, interdisciplinary and dynamic issue, demands to be addressed by multiple stakeholders, who, if they do not work jointly, can fall into individual activism instead of collective construction processes, causing credibility to be lost in the educational-environmental processes.

In cities where there are several universities, it is important to promote interuniversity alliances such as the Manizales-SUMA University System, which leverage substantive missions (teaching, research and social outreach) and allow projecting to the territories, strengthening processes of social transformation.

The intersectoral action promoted by private institutions and non-governmental organizations allows generating spaces for citizen reflection and environmental knowledge management, such as the Environmental Chair presented here. In addition, urban environmental management processes can be promoted, such as the integral management of solid waste in educational institutions, articulating the educational-environmental processes with the technological-logistical aspects required for their effectiveness.

Intersectoral governmental alliances, such as CIDEAMA (Inter-institutional Committee for Environmental Education of Manizales), which are established by Colombian regulations, are participatory planning spaces that mobilize various sector interested stakeholders in building a sustainable city vision. This allows the optimization of resources, the projection of processes over time and the impact improvement of the various actions that each entity undertakes from its institutional purposes.

Acknowledgments

The authors thank the representatives of Suma Ambiental, Fundación Centro de Pensamiento Ambiental, Empresa Metropolitana de Aseo (EMAS), Secretary of Education of Manizales and the member institutions of CIDEAMA for their participation and teamwork to achieve the progress presented.

References

Adomßent, M., Barth, M., Fischer, D., Richter, S. and Rieckmann, M. 2014. Learning to change universities from within: A service-learning perspective on promoting sustainable consumption in higher education. Journal of Cleaner Production, 62(October), 72–81.

Ajzen, I. 1991. The theory of planned behaviour. Organizational Behaviour and Human Decision Processes, 50, 179–211.

Alonso, I. 2012. El impacto de las atribuciones y las normas en el comportamiento anti-ecológico ilegal (Series Doc).

Arnon, S., Orion, N. and Carmi, N. 2015. Environmental literacy components and their promotion by institutions of higher education: an Israeli case study. Environmental Education Research, 21(7), 1029–1055.

Asamblea General NU. 2015. Transformar nuestro mundo: la Agenda 2030 para el Desarrollo Sostenible, 16301, 1–40.

Bamberg, S. and Möser, G. 2007. Twenty years after Hines, Hungerford, and Tomera: A new meta-analysis of psycho-social determinants of pro-environmental behaviour. Journal of Environmental Psychology, 27(1), 14–25.

Calderón-Cuartas, P., Mejía, C. and Vargas, J. 2019. De los valores a las acciones: Una cátedra para promover la producción y el consumo sostenible. (Centro Editorial UCM). Manizales.

Calderón Cuartas, P., Osorio Viana, W., Naranjo Vasco, J. and Guzmán Hernández, T. 2019. Formación de cultura ambiental desde el enfoque de ciclo de vida: una propuesta pedagógica para la sostenibilidad. Ambiente y Desarrollo, 23(44).

Callejas, M., Sáenz, O., Plata, A., Holguin, M. and Mora, W. 2018. El compromiso ambiental de Instituciones de Educación Superior en Colombia. Praxis and Saber. Revista de Investigación y Pedagogía, 9(21), 197–220.

Campos, R. and Camacho, M. 2015. Gestión interuniversitaria y responsabilidad en la gestión ambiental : Plan de acción para el mejoramiento de la gestión integral de los residuos sólidos. Revista Gestión de La Educación, 5, 1–22.

Charpentier, C. 2004. Las barreras para la Educación Ambiental pueden superarse. Revista Biocenosis, 18, 103–108.

Cialdini, Robert and Reno, Raymond and Kallgren, Carl. 1990. A focus theory of normative conduct: recycling the concept of norms to reduce littering in public places. Journal of Personality and Social Psychology, 58, 1015–1026.

Consejo Nacional de Política Económica y Social. and Departamento Nacional de Planeación. Política Nacional para la Gestión Integral de Residuos Sólidos. 2016. Bogotá D.C, República de Colombia.

Consejo Nacional de Política Económica y Social and Departamento Nacional de Planeación. Estrategia para la implementación de los Objetivos de Desarrollo Sostenible ODS en Colombia. 2018. República de Colombia.

Corral-Verdugo, V. 1994. ¿Mapas cognoscitivos o competencias ambientales?.pdf.

Coya, M. 2001. La ambientalización de la Universidad. Un estudio sobre la formación ambiental de los estudiantes de la Universidad de Santiago de Compostela y la Política Ambiental de la Institución. Universidad de Santiago de Compostela.

Décamps, A., Barbat, G., Carteron, J. C., Hands, V. and Parkes, C. 2017. Sulitest: A collaborative initiative to support and assess sustainability literacy in higher education. International Journal of Management Education, 15(2), 138–152.

Dillon, J. 2018. On the convergence between science and environmental education. *In*: Yeo, J., Teo, T. and Tang, K. S. (eds.). Science Education Research and Practice in Asia-Pacific and Beyond. Springer, Singapore.

Emmons, K. 1995. Leaving "behaviour" behind: An alternative. American Perspective on Environmental Action, (1980), 64–77.

Ghazvinei, P., Mir, M., Dravishi, H. and Arriffin, J. 2017. University Campus Solid Waste Management. Combining Life Cycle Assessment and Analytical Hierarchy Process. Cham: Springer International Publishing.

Goldman, D., Assaraf, O. and Shaharabani, D. 2013. Influence of a non-formal environmental education programme on junior high-school students' environmental literacy. International Journal of Science Education, 35(3), 515–545.

Guagnano, G., Stern, P. and Dietz, T. 1995. Influences on attitude-behaviour relationships: A natural experiment with curbside recycling. Environment and Behavior, 27, 699–718.

Ham, S. and Sewing, D. 1988. Barriers to environmental education. Journal of Environmental Education, 19(2), 17–24.

Hines, J., Hungerford, H. and Tomera, A. 1987. Analysis and synthesis o f research on responsible environmental behavior: a meta-analysis. The Journal of Environmental Education, 18(June 2013), 1–8.

Hobbs, L. and Campbell, C. 2018. Growing through partnerships. pp. 139–167. *In*: School-Based Partnerships in Teacher Education: A Research Informed Model for Universities, Schools and Beyond. Springer Singapore.

Hollweg, K., Taylor, J., Bybee, R., Marcinkowski, T., McBeth, W. and Zoido, P. 2011. Developing a framework for assessing environmental literacy. Washington, DC: North American Association for Environmental Education, 122.

IESALC-UNESCO. 2008. Declaración y Plan de Acción de la Conferencia Regional de Educación Superior en América Latina y el Caribe., 73.

Irazábal, C., Mendoza-Arroyo, C., Ortiz, C., Ortiz, R. and Maya, J. 2015. Enabling community-higher education partnerships: Common challenges, multiple perspectives. Current Opinion in Environmental Sustainability, 17(July), 22–29.

Jones, M., Hobbs, L., Kenny, J., Campbell, C., Chittleborough, G., Gilbert, A., Herbert, S. and Redman, C. 2016. Successful university-school partnerships: An interpretive framework to inform partnership practice. Teaching and Teacher Education, 60, 108–120.

Juárez-Nájera, M. 2015. Exploring Sustainable Behavior Structure in Higher Education. A Socio-Psychology Confirmatory Approach. Springer International Publishing Switzerland.

Lizin, S., Dael, M. and Passel, S. 2017. Resources, conservation and recycling battery pack recycling: Behaviour change interventions derived from an integrative theory of planned behaviour study. Resources, Conservation and Recycling, 122, 66–82.

Lopez, F. and Guerrero, J. 2017. Consideraciones ambientales sobre las prácticas de consumo de agua y energía en hogares urbanos. Espacios, 38(59), 28–43.

Mcguire, N. 2015. Environmental education and behavioral change: an identity-based environmental education model. International Journal of Environmental & Science Education, 10(5), 695–715.

Melero, J., Hernández, D., Favela, H. and Ojeda, S. 2013. Actitudes y conductas ambientales en el manejo de residuos sólidos en una IES. In Hacia un sistema de gestión integral de residuos sólidos.

Ministerio de Ambiente, V. y D. T. 2011. Política Nacional de Producción y Consumo. Hacia una cultura de consumo sostenible y transformación productiva. Bogotá D.C.

Ministerio de Ambiente y Desarrollo Sostenible. Política Nacional de Cambio Climático. 2017. Colombia.

Ministerio de Educación Nacional and Ministerio de Educación Nacional. 2002. Política Nacional de Educación Ambiental. Politica Nacional de Educación Ambiental SINA. Bogotá D.C.

Mora, W. 2009. Educación ambiental y educación para el desarrollo sostenible ante la crisis planetaria: demandas a los procesos formativos del profesorado. Tecné, Epistemey Didaxis: Revista de La Facultad., (26), 7–35.

Moreira, C., Araya, F. and Charpentier, C. 2015. Educación ambiental para la conservación del recurso hídrico a partir del análisis estadístico de sus variables. Tecnologia En Marcha, 28(3), 74–85.

Murray, P. and Cotgrave, A. 2007. Sustainability literacy: the future paradigm for construction education. Structural Survey, 25(1), 7–23.

Nelson, J. 2017. Partnerships for sustainable development: Collective action by business, governments and civil society to achieve scale and transform markets.

ONU Medio Ambiente. 2018. Perspectiva de la Gestión de Residuos en América Latina y el Caribe. Ciudad de Panamá.

Osorio Viana, W., Calderón Cuartas, P. and Naranjo Vasco, J. 2019. Strategies for identification of interdisciplinary research subjects: approach, tools, and opportunities for sustainable development of agricultural economies. pp. 275–295. *In*: Pardo Martínez, C., Cotte Poveda, A. and Fletscher Moreno, S. (eds.). Analysis of Science, Technology, and Innovation in Emerging Economies. Palgrave Macmillan, Cham: Springer International Publishing.

Oztekin, C., Teksö, G., Pamuk, S., Sahin, E. and Kilic, D. 2017. Gender perspective on the factors predicting recycling behavior: Implications from the theory of planned behavior. Waste Management, 62, 290–302.

Palacios, J., Bustos, M. and Soler, L. 2015. Factores socioculturales vinculados al comportamiento proambiental en jóvenes Sociocultural factors as predictors of proenvironmental behavior in youths. Revista de Psicología. Universidad de Chile, 24(1), 1–16.

Páramo, P. 2017. Reglas proambientales: una alternativa para disminuir la brecha entre el decir-hacer en la educación ambiental. Suma Psicológica, 24(1), 42–58.

Park, J. and Ha, S. 2014. Understanding consumer recycling behavior: combining the theory of planned behavior and the norm activation model. Family and Consumer Sciences Research Journal, 42(3), 278–291.

Parkin, S., Johnston, A., Buckland, H., Brookes, F. and White, E. 2004. Learning and skills for sustainable development: developing a sustainability literate society. Guidance for Higher Education Institutions, (February), 1–68.

Pauw, J., Gericke, N., Olsson, D. and Berglund, T. 2015. The effectiveness of education for sustainable development. Sustainability (Switzerland), 7(11), 15693–15717.

Programa Manizales Cómo Vamos. 2018. Informe de calidad de vida. Manizales 2018. Manizales.

Programa Manizales Cómo Vamos. 2019. Informe de calidad de vida. Retos con los ODS. Manizales 2019. Manizales.

Rivas, M. 2014. Modelo educativo ambiental para manejo de residuos peligrosos en un institución universitaria de Colombia. Gestión Ambiental, 62(122), 47–62.

Román, Y. 2015. Inclusión de la dimensión ambiental y urbana en algunas instituciones de educación superior en Bogotá. Ambiens. Revista Iberoamericana Universitaria En Ambiente, Sociedad y Sustentabilidad, 1(2), 139–160.

Roth, C. 1992. Environmental literacy: Its roots, evolution and direction in the 1990s. Education Development Center, 51.

Sáenz, Orlando and Benayas, J. 2015. Ambiente Y Sustentabilidad En Las Instituciones De Educación Superior En América Latina Y El Caribe*. Revista Iberoamericana Universitaria En Ambiente, Sociedad Y Sustentabilidad, 1(2), 192–224.

Sáenz, O., Plata, A., Holguín, M., Mora, W. and Blanco, N. 2017. Institucionalización del compromiso ambiental de las universidades colombianas. Civilizar Ciencias Sociales y Humanas, 33(July), 189–207.

Saer, A. and Romano, C. 2019. Estrategia Nacional de Economía Circular - ENEC.

SISPD y DNP. 2017. Disposición final de residuos sólidos. Informe Nacional 2015, 140 páginas.

Steffen, W., Broadgate, W., Deutsch, L., Gaffney, O. and Ludwig, C. 2015. The trajectory of the anthropocene: The great acceleration. Anthropocene Review, 2(1), 81–98.

Stibbe, A. (ed.). 2009. The Handbook of Sustainability Literacy: skills for a changing world. Green Books, 1–6.

Schwartz, S. H. 1977. Normative influence on altruism. pp. 221–279. *In*: Berkowitz, L. (ed.). Advances in Experimental Social Psychology, Vol. 10. New York: Academic Press.

Thomas, C. and Sharp, V. 2013. Understanding the normalisation of recycling behaviour and its implications for other pro-environmental behaviours: A review of social norms and recycling. Resources, Conservation and Recycling, 79, 11–20.

Tonglet, M., Phillips, P. and Read, A. 2004. Using the theory of planned behaviour to investigate the determinants of recycling behaviour: A case study from Brixworth, UK. Resources, Conservation and Recycling.

UI Greenmetric. 2020, marzo. List of Universities in Each Country (2019) http://greenmetric.ui.ac.id/country-list2019/?country=Colombia.

ULSF. 1990. Declaracion de Líderes de Universidades para un Futuro Sostenible Declaración de Talloires. Retrieved from http://ulsf.org/wp-content/uploads/2015/06/Spanish_TD.pdf.

UNEP. 2017. Consuming Differently, Consuming Sustainably: Behavioural Insights for Policymaking. The United Environment Programme.

Unesco-Pnuma. 1978. Conferencia intergubernamental sobre educación ambiental. Tbilisi (URSS). Octubre. Informe Final.

Vallejo, G. 2013. Reflexiones, elementos y perspectivas para pensar la dimensión ambiental en las instituciones de Educación Superior en Colombia (PRAU). Universidad Libre.

Vicente, M. A., Fernández, A. and Izagirre, J. 2013. Environmental knowledge and other variables affecting pro-environmental behaviour: Comparison of university students from emerging and advanced countries. Journal of Cleaner Production, 61, 130–138.

WWF International. 2018. Informe Planeta Vivo 2018: Apuntando más alto. Gland, Suiza.

CHAPTER 9

Successful National Financial and Industrial Strategies for the Adoption of New Renewable Energies and the Creation of Exporting Companies

Jaime Torres-González

Introduction

Some nations are more proactive and visionary in establishing regulatory frameworks and incentives that encourage the creation of companies and complete productive sectors. This stimulates the early development of technological innovation. One of the areas of human activity that requires urgent transformation is the generation of energy, given its effects on global warming. The threat of the growing pollution of the planet exists due to the current fossil fuels, which means a palpable risk for both human societies and ecosystems. This crisis requires urgent responses in terms of replacing fossil fuels and developing new energy alternatives.

Despite the progress achieved so far and the agreement signed at COP21 in Paris in 2015, it is clear that the replacement of the polluting fossil matrix is still very slow (IEA 2019, UN Intergovernmental Panel on Climate Change 2018). It indicates that the risk of the sustainability of human activity on the planet is growing and that successful schemes for the adoption of new renewable energies (NRE) should be expanded on a much larger scale and speed in all countries. The direct interventions explained by the young Swedish girl, Greta Thunberg at the UN COP24 on Climate change in Poland 2019 should be seriously considered:

> "Our civilization is being sacrificed for the opportunity of a very small number of people to continue making enormous amounts of money. Our biosphere is

University of Bogotá Jorge Tadeo Lozano International Trade and Finance Program.
Email: Jaime.torresg@utadeo.edu.co

> being sacrificed so that rich people in countries like mine can live in luxury … Until you start focusing on what needs to be done rather than what is politically possible, there is no hope. We cannot solve a crisis without treating it as a crisis. We need to keep fossil fuels in the ground, and we need to focus on equity. And if solutions within the system are so impossible to find, maybe we should change the system itself … We have run out of excuses and we are running out of time."

This research is organized to first look at the structure of the current global energy matrix and analyze its strong fossil component. It is shown that some countries use the most energy, leading to high polluting implications and that the share of renewable energies is still very low. The following sections analyze the main policies, strategies and mechanisms that several countries, including Germany, China and India, have developed to successfully increase their generation of renewables. The objective is to understand the main elements of the political economy that have allowed them to advance above average. In addition, analyze the cases of countries that have begun to generate significant amounts of NRE, such as Mexico and Costa Rica, and finally Colombia, a country that has recently explored new renewables. The work is concluded by considering the industrial and technological policies that two leaders, Germany and China, have implemented.

The Dominant Fossil Energy Matrix and the Incursion of Renewables

In Figure 1, it can be seen that the main energy resources of the twenty-first century continue to cause a lot of pollution, such as coal, oil and natural gas, while renewables are still marginal. The consequence was CO_2e pollution grew by 64% in this century, from 23 Gt CO_2e in 2000 to 32.8 Gt in 2017 (International Energy Association [IEA] 2019a). Provisional data for 2018 indicate that they grew faster that year than in 2017 due to much of the economic growth in Asian countries being supported by coal-fired, thermal and industrial plants. IEA projections indicate that pollutant emissions

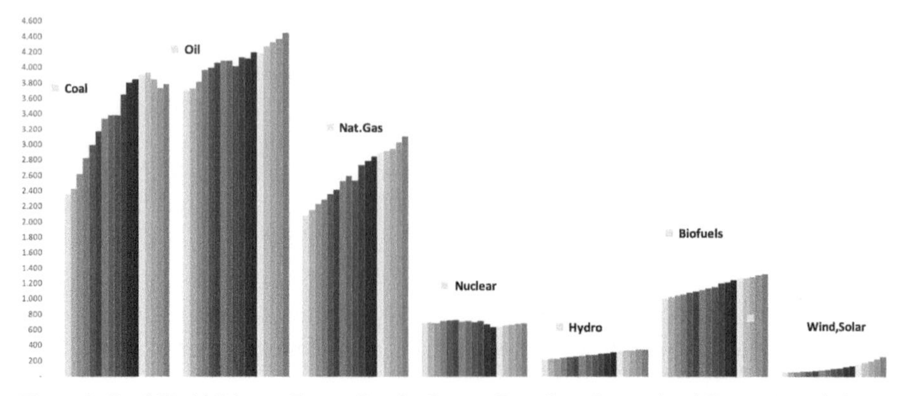

Figure 1. Total World Primary Energy Supply. Source: Data from International Energy Association – IEA, World Energy Balances 2019.

will continue to rise at least until 2040 (IEA 2019b) with serious effects on global warming.

Some developed OECD member countries have begun to reduce their contributions to global pollution in recent years in terms of Total Primary Energy Supply (TPES) (Figure 2) by starting complex energy diversification plans and listening to the social pressures demanded by new environmental policies. However, their reduction is not yet decisive as they continue to use fossil fuels massively. The picture is grimmer for most developing nations, which are increasing their consumption of all primary energies (coal, oil and gas) and electricity.

It can be seen in Figure 2 that countries such as the USA, Japan and Germany reduced TPES by the end of 2010. Others such as Mexico, Chile and Colombia stabilized, while the most dynamic emerging countries due to their economic growth, such as China, India, Korea, Indonesia, South Africa and many other countries, rapidly increased their total energy consumption (OECD 2019).

In Figure 3, it is illustrated how in countries such as Germany between 2000 and 2017, the energy generation with renewable sources grew at a rate of 9.7% annually. If this trend continues, it would allow a faster replacement of contaminating sources. In Costa Rica, they grew at 5.2% until 2016, which makes it one of the few nations on the planet that generates about half of its TPES with renewable resources (see Figure 4). While in Colombia, Chile and the USA, the renewables increased

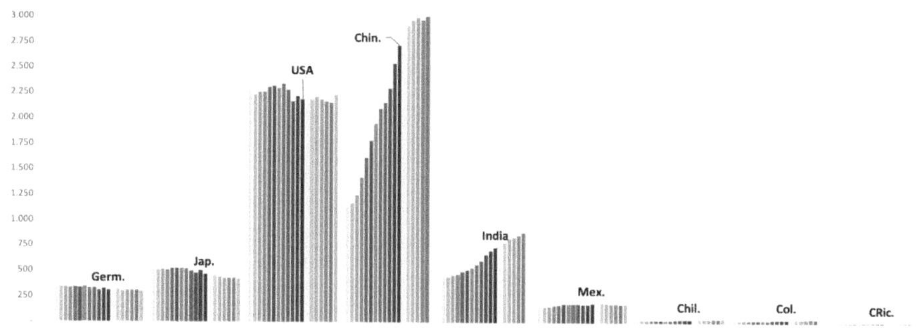

Figure 2. World Primary Energy Supply. Source: OECD, World Energy Statistics 2019. (Germ: Germany, Jap: Japan, Chin: China, Mex: Mexico, Chil: Chile; Col: Colombia; Cric: Costa Rica.

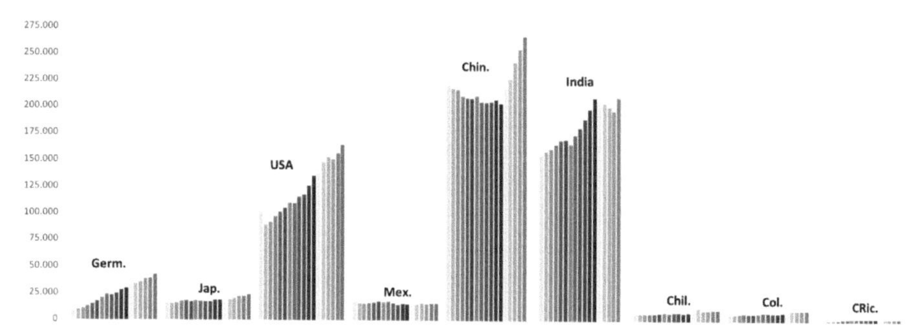

Figure 3. Renewable Energy Supply. Source: OECD, World Energy Statistics 2019.

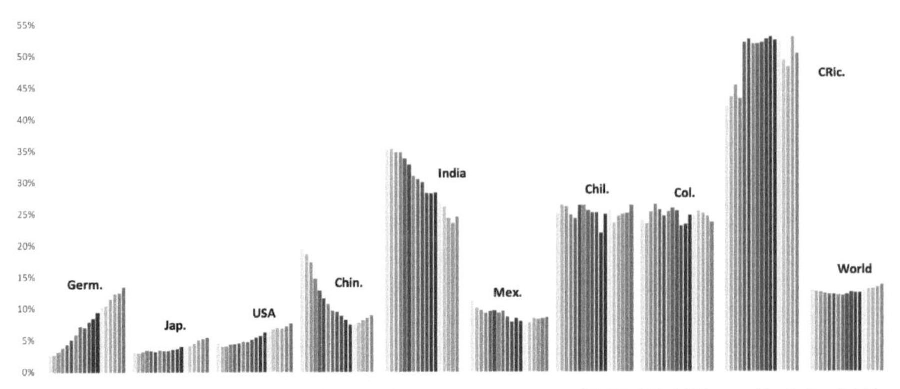

Figure 4. Renewable Energy Share of Total Primary Energy. Source: OECD, World Energy Statistics, 2019.

on average by 3.0%. This is not a negligible rate, but it would only allow a slow substitution, and it is less than needed with the urgency of climate change.

Energy generation with water decreased in China between 2000 and 2011 (Figures 3 and 4) to 203 million Toe (OECD 2019) but as of 2012 this trend for renewables was reversed due to its aggressive wind and solar installation programs, growing in this new phase at an average annual rate of 5.6%.

In another highly populated nation such as India, the growth of renewables indicated by the OECD 2019 data show that they increased at a moderate annual rate of 2.7% between 2000 and 2011, decelerating by 1.5% annually until 2015 and recovering in 2016 when the government began to stimulate the adoption of solar technologies. However, due to its rapid economic growth and the great contribution of coal, the share of renewables decreased (see Figure 4). Additionally, part of its generation classified as 'renewable' includes a high use of biomass that corresponds to the use of wood for cooking, which generates CO_2 and induces negative impacts on the health of users and the environment by deforestation (EPRS 2018).

Figure 4 shows the share of renewable energy of the total energy used by countries. The results are mixed; in nations such as Costa Rica, renewables make up half of the total and in Chile and Colombia a quarter, all with relatively stable trends. In India, the participation of renewables has rapidly decreased because the use of traditional pollutants, especially coal, has increased very rapidly in the twenty-first century. In China, the share of renewables (especially hydro) declined steadily from the year 2000 (19.5% of the total) to the year 2011 when it reached a minimum of 7.5%. This is mainly because consumption grew faster than the new renewable capacities China installed. The participation of renewables recovered and in 2016 it accounted for 9% of the total used. Similarly, in many other developing nations, there is a rapid economic and population growth as well as the structural transformation of their societies toward more urban ways of life with greater comforts and large industries. This makes their per-capita consumption rise and the share of renewables decrease. All of this shines a light on the necessity to reduce climate deterioration.

In more developed countries, such as Germany, Japan and the USA, the share of renewable energies with respect to the total energy exhibits only marginal gains; in 2017, their share in Germany was 13.6%, in Japan only 5.4%, while in the USA a scarce 7.6%. The average of the world had its minimum in 2007 with a 12% participation and shows a soft tendency to the growth of renewables, reaching 13.7% in 2016. It can be concluded that the renewable energy share of the total energy matrix is still just a marginal gain. Its growing trend is only modest in relation to the sustainable development commitments of the COP21 and the need for replacing sources of pollution to prevent climate change.

The nations analyzed in this study are a good sample of the planet and exhibit as a characteristic that those who initiate a per-capita reduction of their CO_2e generation, still provide a high pollutant volume. Germany doubles—and the USA quadruples—a global average of CO_2 production because of high energy and materials consumption (see Figure 5). But most developing countries, both small and large, despite presenting lower levels of per-capita pollution, increase them steadily in order to improve their productive capacities using the available technologies, which are intensive in the use of fossil fuels.

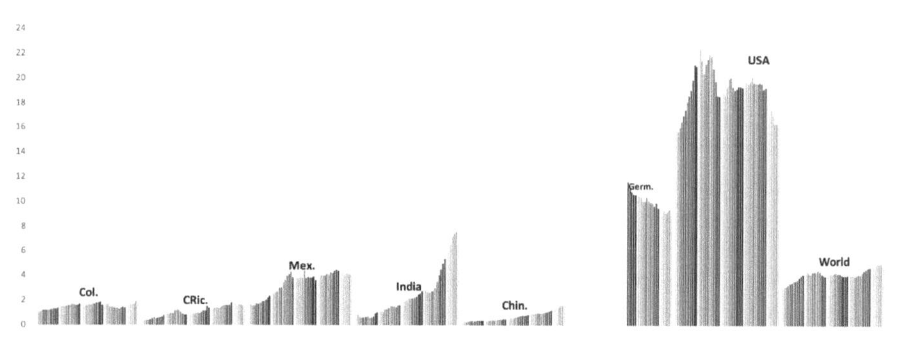

Figure 5. CO_2e emissions per-capita. Source: World Bank – World Development Indicators.

To carry out a more detailed analysis, cases of representative countries will be examined for recent trends in their NRE adoption. It will be illustrated how they can lead a counter approach against the development of polluted industrial procedures. Specifically, it is vital to investigate the primary energy sources each country uses, their use of NREs and electrical energy.[1] There are countries that have been successful in the primary phase of the development of NRE, such as Germany. There are other very populated nations that a few years ago had decided on ambitious replacement programs, such as China and India, as well as other countries in Latin America that have recently adopted policies that stimulate the rapid adoption of NRE, such as Mexico and Costa Rica and Colombia which is just beginning this path.

For all countries, the evolution of the regulatory frameworks that have sheltered the renewable energy sector will be analyzed. The main characteristics and instruments that have favored the adoption of NRE will be identified and their achievements through the installation of infrastructure and the reduction of their levels of environmental pollution will be laid out. Some countries not only adopt these technologies but deliberately define industrial policies that seek to position

[1] Electricity can be generated either with fossil fuels or with renewables.

them as leaders in the production of NRE. One of these countries is Germany, whose experience we review succinctly.

The "Green Policy" in Germany

This nation, similar to all those with large industrial capacities, was strongly affected by the oil crisis of the 1970s when its supply was interrupted and the price of crude oil became very expensive. Its great dependence on fossil fuels was illustrated at that time and the government became determined to support the research of other energies, including nuclear and new renewables. This would help overcome its fragility (García-Alvarez and Mariz-Pérez 2012). Different research institutions intensified trials with old technologies, such as windmills and biomass energy, and new options, such as solar energy, photovoltaic, as well as nuclear energy, which already had strong citizen opposition during the 'cold war', which intensified with the serious accident in Chernobyl (Ukraine, USSR) in 1986.

This complex set of realities gave rise in Germany to the birth of the "Green Movement" (Die Gruene) that marked the protest against growing environmental pollution and nuclear risk. It also led to the analysis of alternatives for energy self-generation, pushing German politicians to be more sensitive and responsible for the challenges of climate change. This background enabled support in the public realm to develop a new policy framework, which through a paradigm shift and large investments, formed a pioneering public policy that was issued in 2000 (Erneuerbare-Energien-Gesetz [EEG]). This is now counted among the most successful frameworks for the massive adoption of renewable energies.

The main political and strategic objectives that Germany seeks to achieve (Wassermann et al. 2012, Waffenschmidt 2017, Fraunhofer ISE 2018) are:

- Reduction of dependence on coal, petroleum products and nuclear energy
- Reduction of pollution with greenhouse gases
- Diversification of energy sources
- Reduction of political risks due to instability of oil-producing countries
- Scientific leadership for the creation of environmental technology
- Job creation
- Founding of NRE manufacturing and exporting companies.

To help build an enabling environment, its regulatory framework was based on the following principles:

- State grants for NRE research and development (first phase)
- Private subsidy paid by consumers through electricity tariffs to finance the higher costs of electricity generation with NRE (Second phase, from 2004 to 2023/4).
- The state forced traditional electricity providers to purchase (or produce themselves) defined quotas of electricity from renewable sources, opening the basket for new sources.[2]

[2] One example was the preferential energy purchase to the 100.000 roofs program investors (Renewable Energy Law-2000), goal that was reached in 2003 and led to a rapid increase in photovoltaic capacity.

- Creation of the "Feed-in-tariff" system (Stromeinspeisungsgesetz 1991), which granted financial and legal stability so that new investors in decentralized renewable energy systems could generate, connect to electricity networks and sell their surpluses. This system enabled 20-year contracts with stable prices for sales of kWh, so that the new generators kept their income safe during the term of the investment without risk due to technological, exchange rate or tariffs change.
- Creation of new generators – people, companies and municipalities through the incentive to invest in decentralized systems ("Buerger Energiewende"), such as windmills, biodigesters and solar panels, with a legal obligation to be incorporated into existing grid networks by big traditional distributors.
- The last phase of the German model consists of auctions. These are regulated by the government since 2016, and companies offer electricity at auction prices, eliminating the need for subsidies.

It is today well known that thanks to recent technological advancements, current costs of NREs are persistently reducing, and they are every new day more competitive with all fossil fuels (IRENA 2019). This is possible even in countries with lower levels of solar radiation, such as Germany (World Bank Group 2020). The modality of auctions favors the participation of medium and large investors, which are in a better position to reduce costs due to the use of larger generation scales.

The determined stimulus packages for NREs allowed Germany to become a pioneer country and set an example for others. The success of its policy can be seen in indicators that show how the annual generation of electric power with wind power rose from 7.5 TWh in 2000 to 77.4 TWh in 2016 (10-fold increase). The photovoltaic electric power rose sharply from almost zero (0.06 TWh) in 2000 to 38.2 TWh in 2016, and biofuel production improved from 22.9 TWh in 2000 to 371.9 TWh in 2016 (16 times) (BP Statistical Review 2018). Germany was placed globally as the fourth largest investor in NRE in the period from 2020 to 2019 first quarter with US$ 179 billion, after China, the USA and Japan (FS et al. 2019).

Germany managed to reduce its greenhouse gas emissions (in its CO_2e) from its historical maximum of 1,400 Million tons in 1979 to 866 Mt in 2018. However, it is seen today that it would not be easy to comply with the commitment acquired at COP21 to reach 751 Mt in 2020 (40% below its 1990 emissions). Germany is today generating just about 16% of its total energy consumption with renewable energy (Clean Energy Wire 2019), indicating that to meet its goal of reducing 95% of pollutant emissions by 2050 (generating only 63 Mt CO_2e), it will still have to take more decisive actions than it is currently.

So, to say, one of the most advanced large countries on the globe, which is a leader in the generation of non-polluting energies, still requires 84% of traditional polluting resources to meet its energy consumption today (data for 2018). This indicates how far we are from an actual replacement of fossil fuels. On a visit to China by the German Minister of the Environment, Ms. Svenja Schulze on October 29, 2019, said at her embassy in Beijing: "The international community goes so slowly in order to achieve the compromises of the Paris Agreement".

China: The Largest Absolute Pollutant and the (Recently) Largest Investor in NRE

China needs to import about one-fifth of its energy resources (CREO 2017). This depends on large purchases of oil and derivatives, a resource susceptible to high geopolitical and financial risks. The need for these resources has grown in China as in no other nation on the globe in recent decades. Of the total energy used in the world in 1975, China participated in only 7%. By 2016, it had risen to 22% (2,958 Mtoe), far exceeding the USA (IEA 2019). For its supply, this country relies heavily on the exploitation of coal (3,900 million tons annually), the most polluting fossil fuel. The implications within China are clear; different studies show that in 2013, seven of the 10 most polluted cities on the globe were in China, and about 1.6 million people died each year from respiratory diseases (Chiu 2018).

Figure 5 shows the rapid growth of the per-capita emission of pollutants in China. This giant went from emitting 3.2 Gt CO_2e in 2000 to 9.3 Gt in 2017 (IEA 2019), almost tripling its emitting power in 17 years, in absolute terms. It should be noted that the Chinese efforts to reduce its CO_2 emitting volume meant that these gases lost participation as a percentage of their GDP. CO_2 emissions fell from 4.4% in 2000 to 3.2% in 2017 as a proportion of its economic activity (IEA 2019). Due to the greater perceived risk and the over-reliance on coal generation in the country, the government decided to reduce the installation of new coal-based power plants in 2016. This policy decision led to the cancellation of 15 approved plants with 12.4 GW capacity and stopped the approval and construction of new coal units in 28 of the country's 32 provinces (CREO 2017).

Despite these actions, it can be seen that while most developed countries began to lower their absolute CO_2e emissions a few years ago (Figure 6), China and emerging countries did not. Their first steps had been to replace the more polluting resources and improve energy efficiency, producing with less energy, but their absolute volume of pollutants emitted is still increasing, making the effects of global warming worse.

China's incursion in the generation of NRE is relatively new, and these have become a high-value political objective, expressed in the importance assigned to it

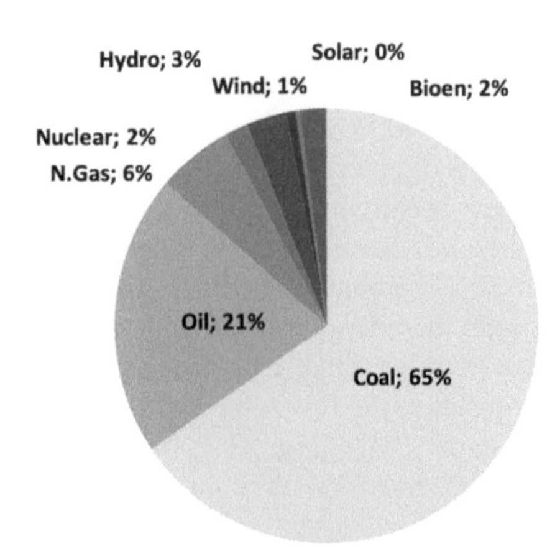

Figure 6. China TPES 2016.

in its five-year development plans (CREO 2017). Its investment meant that between 2010 and 2019-1 it became the largest NRE investor in the globe with US$ 758 billion, doubling the amount of the USA (FS et al. 2019). It can be stated that of the six 'political objectives' defined by Germany (section 3) in its development of NRE, all are applied by China. Although without a doubt, pollution reduction is much more urgent in this nation. The expected results of the regulatory and incentive frameworks that the Chinese state is deploying can be seen in the following projections.

Since China is currently the country that produces the greatest absolute pollution (the USA being the largest per-capita major pollutant) and its CO_2 emission volume will continue to rise at least until 2030, it is shown in Figure 7 a projection of total use of primary energy by 2050 (projection with a goal to avoid global warming greater than 2°C degrees) carried out by the China National Renewable Energy Center (CNREC) (CREO 2017).

As shown in Figure 6 for 2016, coal constituted for 65.2% of the primary energy used by this country. If other fossils and nuclear energy are added, this supply amounted to 93.5%. That is, NREs, including hydro, only contributed 6.5% of the total. If the projection for 2050 is fulfilled (See Figure 7), this scenario would change radically with coal providing only 13.6%, total fossil and nuclear energies would participate with 42.7%, and renewables would amount to 57.3%. To meet these goals, it is necessary to continue rapidly replacing coal with electricity generation by wind, solar and bioenergy, which will form the basis of its future supply.

This implies that they must continue eliminating coal-fired power plants and develop a policy to control the use of fuelled vehicles as well as industrial and residential equipment propelled by oil and gas. This requires a strong adoption of electric vehicles and stimulating industrial and urban energy efficiency policies, which would be more likely to be achieved with the wide use of the internet and 5G technology. Wind and solar would be the biggest new contributors to the energy matrix, while oil and its derivatives would occupy the second place due to the dependence on fossil fuel use in transport. Coal would see its contribution fall from 1,981 Mtoe in 2016 to 363 Mtoe in 2050 (81.7% reduction) in a great effort to decarbonize the Chinese economy.

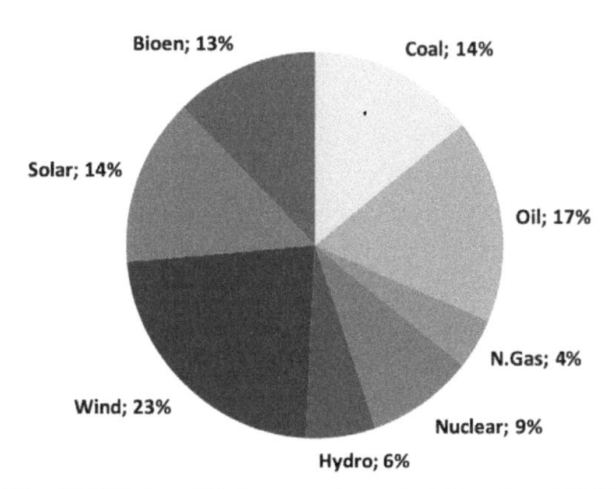

Figure 7. China TPES below 2°C. Sources: Own calculations from CREO 2017: 152.

Although this is an ambitious projection, the recent achievements of this country in terms of NREs are very good and to date, its policies, incentives and investments have allowed it to surpass its defined goals. The use of a feed-in-tariff system, the purchase of renewable energy quotas defined by the state, banks financing actively NRE projects and lastly, energy auctions for NREs; all these measures have reported important achievements.

Energy Consumption in India, the 2nd Most Populous Country on the Globe

Figure 5 shows how India still has a low per-capita impact on its CO_2e emissions, but due to its rapid economic growth in the twenty-first century (6.5% annual average, BP 2018) and its large population, there has been an improvement in the quality of life that requires the use of more energy. Therefore, India has become the third-largest polluter on the planet, exceeding 2 Gt CO_2e in 2016 (similar to China's emissions in 1990, IEA 2019) and it is very likely to occupy the second place soon. It is estimated that in 2014 there were 270 million people in the Indian countryside without electricity. In September 2018, the government announced that 91% of them had been connected to the service. However, a huge number (780 million) still use biomass (wood, coal or animal dung) for cooking (European Parliament 2018).

The low country's self-sufficiency in the primary energy generation constitutes a fragile point of its current economic structure with large volumes of oil and natural gas being imported. In 1990, it imported 15% of national energy consumption, in 2012 it was 38%. In 2016, the country met 93% of its transportation and industrial energy needs with fossil fuels and it is estimated that by 2040 it will still be 82% (EPRS 2018).

The BP – World Energy Outlook 2040 (BP 2018: 55) says: "In the Energy Transition scenario, China's coal intensity declines sharply, with its overall coal consumption falling, more than offset by a large rise in renewable energy. Indeed, the largest growth of any energy source at a regional level is the increase in renewables in China. In contrast, the share of coal within India and other emerging Asian countries is largely unchanged, such that coal demand increases along with overall energy demand."

To avoid a scenario so dependent on fossil fuels and to improve the standard of living of its large rural population, India has recently embarked on an active NRE stimulus and investment policy. It is estimated that the potential solar energy capacity of the country is 750 GW of which up to 2018, only 32 GW had been used, that is, 4.3% (FS et al. 2019). However, its government established commitments at the COP21 in Paris to install 175 GW of NRE in 2022 with 100 GW solar energy and 60 GW wind energy (IBEF 2018), which marks the beginning of a trend toward greater use of renewable resources. Between 2010 and 2019-1, US$ 90 billion was invested in the NRE sector (FS et al. 2019), placing this country as the sixth-largest investor in the globe. To stimulate the investments of the 2018–2022 Program in this sector, the government has granted the following benefits (IBEF 2018):

- Zero import duty on capital equipment and raw materials
- Low-interest rates and priority sector lending

- Tax holiday of 10 years for offshore wind energy generation
- State Electricity Commissions are obliged to purchase a percentage of power from renewable energy providers under Renewable Purchase Obligations. It is also "floor price" guaranteed (US Cent 14.4/kWh in 2018)
- Single window mechanism is provided for all permissions for NRE projects
- A 100% Foreign Direct Investment is allowed under the automatic route for NRE generation and distribution (Electricity Act 2003).

It can be concluded that India has discovered a huge source of renewable energy, especially solar energy, and has recently started a way to develop them. It is becoming one of the largest nations on the globe with faster investments in NRE. In March 2018, it opened the largest solar park in the world (Shakti Sthala) with 2 GW of installed capacity and investment of US$ 2.55 billion in the dry Karnataka region. Further, India has scaled up the target of the National Solar Mission to 100 GW by 2022 from 20 GW of grid-connected solar power by 2018 (IBEF 2018).

In just two years (2017–18), the country added 11.8 GW of renewable energy to its installed capacity (similar to the 2018 total installed water capacity in Colombia), indicating a great national commitment to the development of the NRE. However, due to its enormous energy needs to satisfy about 1.4 billion inhabitants and its rapid economic growth, it can be seen that the country is just at the beginning of its energy transition. Nevertheless, the most used resource will continue to be coal, implying that its contribution to global pollution will continue to grow for several decades. In other words, the NREs need much greater stimuli and investments than those made so far.

New Renewable Energy Developments in Mexico

As can be seen in Figure 5, the per capita volume of CO_2e emission in Mexico is similar to the global average. However, its renewable energy generation had been stagnant, from 16.9 MToe in 2000 it dropped to 15.6 Mtoe in 2017 (OECD 2019, see Figure 3), thus increasing the share of polluting energies in its total energy matrix (Figure 4). Considering all of that, this country still has an excellent level of solar irradiation and areas of great potential for solar installations (World Bank Group 2020). To use them, the Mexican government called three auctions between March 2016 and November 2017, approving 15- and 20-years contracts for the installation of 25 wind farms with the addition of 1,834 MW and 40 photovoltaic parks with 5,057 new MW, totaling an additional 6,891 MW and investments for about US $ 8.6 billion between 2018 and 2020.

It should be noted that not only did the auctions allow multiplying the installation of wind and photovoltaic parks in Mexico for 2018–20 (see Figure 8), but they also lowered the average prices at each auction, going from US$ Cent 4.8/kWh in the first auction in March 2016 to 3.3 in the second and 2.1 in the third in Nov. 2017. This became the second-lowest average price in the globe, after the Saudi Arabian auction in 2018. The fourth auction to be held in 2018 was suspended by the new government, which has not yet defined its renewable energy policy.

With the boost that renewable energy generation has had, the Mexican Solar Energy Association reports that the important photovoltaic additions made in 2018

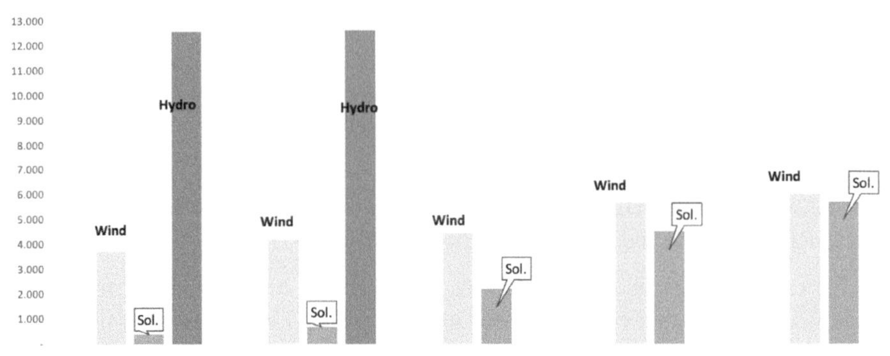

Figure 8. Mexico renewable energy capacities. Source: SENER, México – Reporte de Avance de Energías Limpias, Jun. 2018.

raised the operating capacity by 32% and up to 30 June 2019 had 4,057 MW installed in the country, of which 693 MW (17%) corresponded to 94,893 residential and commercial solar roofs. The goals set by the country for COP21 were the production of 25% of electricity with renewable energy in December 2018, 30% in 2021 and 35% in 2024 (SENER 2018).

The main incentives offered by Mexican legislation are found in article 53 of the Electricity Industry Law of 2014 with the following advantages for its auction policy:

- Contracts for 15 and 20 years, which give legal and price stability to investors
- Projects near consumption centers are rewarded, reducing costs and losses due to transmission
- Auctions do not distinguish between technologies, enabling competition between them and the reduction of prices offered
- Each auction competes not only for installed capacity and generated volume but also for the rights to 'Clean Energy Certificates – CEL', which add profitability to investors and stimulate the greater use of NRE in the country
- The use of net-metering, net-billing and the sale of surpluses of small distributed photovoltaic producers was approved in 2017 by the Energy Regulation Commission-CRE, enabling the massive installation of solar roofs throughout the country.

For the 2010–2019-1 period in Mexico, US$ 23 billion had been invested in NRE (FS et al. 2019), placing the country at Nr. 14 among the largest investors in the world. With recent investments, Mexico ranked as the fifth largest investor in photovoltaic energy in 2018 (Mexico Photovoltaic 2019). Regarding the massification of distributed photovoltaic systems, Mexico`s D.F. Secretariat of Economic Development in cooperation with the Inter-American Development Bank (IDB), 2019, calculated that Mexico City has a potential to cover 48% of its electricity needs with solar energy. This amount is equivalent to 6.0 TWh/year, and the city could replace 38% of the LPG[3] used to heat water using solar heaters. The total investment to be made would amount to US$ 7.44 billion (IDB 2019).

Given that close to 70% of the Mexican territory is suitable for distributed photovoltaic installation, a program such as the one proposed for D.F. could be extended nationally and replace a large part of the state subsidy to residential users of electric power (99% of users). This subsidy represented between 2004 and 2017 a public expenditure of Mexican $ 1,350,000 million at present value (US$ 71 billion). This value could be invested in distributed photovoltaic systems, which would mean saving on public spending and subsequent self-sustainable generation with minimum cost. It is estimated that the cost of the electric subsidy has represented between 0.8% and 1.3% of the country's GNP in recent years (CEFP 2017). This amount could be saved if the photovoltaic expansion is strongly supported in the country, contributing additionally to reduce environmental pollution and improving air quality in cities.

Costa Rica: 100% Green Electricity Generation

In Figure 4, it is depicted how this small Central American country supplies its primary energy needs using about 50% with renewable generation based on geothermal, hydro, wind and biomass, giving an example to the rest of the world on how an energy matrix can be structured with few pollutants. "On March 25, 2015, Costa Rica announced that it would have produced, for 75 consecutive days, 100% of its electricity from renewable sources" (González-Jiménez 2016).

Authors, such as González-Jiménez (2016), consider that this country's achievement in terms of sustainability is based on the historical development of a citizen's vision of the state, which has managed to involve its population in the definitions of energy policy since the early twentieth century. Based on its comparative advantages that allowed the use of the geothermal resource, hydro and biomass to generate electricity at the first stage and subsequently the incorporation of new technologies with minimal environmental impact (like wind). This country has managed to structure an institutional and regulatory framework to give privilege to the environmental sustainability of its electricity generation.

In 1928, and after recovering the democratic institutionality after a military coup, the "Nationalization of the Electric Forces' Law" was issued and the National Electricity Service (SNE in Spanish) was created. This sought as central objectives to democratize access and the definition of energy policy through state control, which would prevent private or foreign monopolies' formation. This strategy led (1945) to the creation of the Association for the Defence of the Electric Consumer and the Costa Rican Institute of Electricity (ICE) in 1949. The financial consolidation of ICE was made possible by state support and its permission to issue bonds and contract loans with national and central American banks in the early 1960s. The work was complemented by the constitution of rural electrification cooperatives (González-Jiménez 2016), which allowed to bring electricity and more democratic economic development to the country. As a result, Costa Rica's 2015 electricity generation

[3] LPG: liquified petroleum gas.

matrix was composed of 73.3% hydropower, 13.1% geothermal, 10.9% wind and 2.7% others (Segura 2015).

The determination of the country's broad social layers to define the electric service as a 'public service' provided and regulated by the state with space for non-monopolistic private investment has allowed Costa Rica to avoid a characteristic structure of the energy matrix in almost the entire planet. This is featured by the use of the cheapest energy sources (coal and diesel) without taking into account their environmental impacts. Decisions taken mainly under this realm (low cost) give priority to a convenient short-sighted financial benefit/cost ratio, involving none environmental considerations.

This type of poor policy was reinforced in the world during the last decades with the competitive liberalization strategy for the electric sectors promoted by multilateral financial institutions (Washington Consensus), which financed the adoption of a large number of investments with fossil fuels. The resulting energy matrix is judged today as harmful to the environment and ends being expensive when the high environmental costs involved are internalized. The reversal of these policies can be seen in the broad financing that the InterAmerican Development Bank (IDB) is currently giving to the deployment of NREs in Latin America.

The conversion from fossil fuel generation presents serious economic and financial conflicts because some expensive coal-electric plants in China, India or Australia or nuclear power plants in Germany or Japan must be closed due to their great environmental impacts. These are now not cost-efficient in the presence of taxes on coal and polluting resources. These plants are no longer competitive with technological innovations on wind and solar, which have substantially reduced their installation and operational costs. The replacement costs are heavy, not only for big countries but also for small ones. A transition period would be needed here. But the inability to quickly disassemble said fossil fuels can mean a delay on the necessary installation of NRE. Fossil fuel plant owners' argue loss of profits associated with stopping them.

This contradiction between private profit and social benefits is slowing the fastest installation and generation with clean sources, slowing down the decarbonization of our societies to avoid the rapid environmental degradation in progress.

In Costa Rica, and despite its obvious sustainable energy achievements (Zárate and Ramírez 2016), 63% of the total energy consumption comes from oil and derivatives, preferably used by the transport sector. Oil must be fully imported, generating an annual expenditure of close to US$ 2 billion between 2011 and 2014 for the importation of 18 to 19 million barrels. The replacement of these fossil fuels will require improving efficiency and reducing the size of vehicles, a solution that will mean the introduction of electric vehicles in the near future.

To continue its efforts, the government defined in 2015 the axes of the VII National Energy Plan of Costa Rica for 2015–2030 (MINAE 2015) with the following objectives:

- Energy sustainability and diversity with low emission levels
- Harmonization of conflicting interests between efficiency, competitiveness, and environmental sustainability

- Stimulus to distributed power generation and small production
- Private and public transport more efficient and environmentally friendly
- Costa Rica's goal is to become carbon neutral by 2021

The regulation of Costa Rica (IEA 2019c) exempts equipment used for renewable energy from:

- Import duties
- Other tax exemptions: excise tax, ad valorem tax, general sales tax, and specific customs tax. Tax exemptions cover equipment for PV panels, solar water heaters, wind and hydro equipment for private electricity generation.

Other mechanisms to support clean technologies:

- It was adopted 'net mettering' as the tariff method for distributed energy measurement.
- Consumers (> 240 MWh/year) are to develop energy plans that include energy efficiency and renewable energy options.

For the definition of the VII Plan, the Ministry of Environment arranged an extensive program of consultations with citizens, trade unions, cooperatives of power generators, professional associations and political parties (MINAE 2015). This behavior continues its tradition of population involvement, respect for sustainability objectives, access by non-monopolistic private investors and keep on developing the country's natural comparative advantages.

Colombia: New Generation with NRE

Colombia has developed a primary energy production matrix strongly based on fossil fuels, i.e., coal, crude oil, and natural gas, accounted for 94% of its total primary energy production in 2015 (141,990 ktoe); 65% of them were exported (see Figure 9). Of the non-exported production (49,729 ktoe), only 29,655 ktoe were consumed locally (59.6%). The difference means that 20,074 ktoe (or 40% of the availability after exports) were missed or "cannot be explained", according to the National Department of Planning (DNP 2017). The not-found fossil fuel could be explained due to technical and non-technical losses, among which would be measurement deficiencies, network thefts, guerrilla attacks on pipelines and corruption in distribution. So, the country has a production matrix that is highly CO_2 generating and a very large amount of these resources are lost. On the other hand, the participation of renewables in the primary energy matrix is still a minority (6% of total supply). Taking into account only local consumption (see Figure 10), 16% corresponded to renewable energies and 84% to polluting resources.

Of the electric power generation, about 70% was produced with water by reservoirs with an installed capacity of 11.8 GW in 2018. The other 30% of electrical energy is generated with fossil resources. It can be concluded that the comparative advantages that Colombia has for being located at the equator and having high solar radiation and is practically yet to be used until 2018. There was only one wind farm

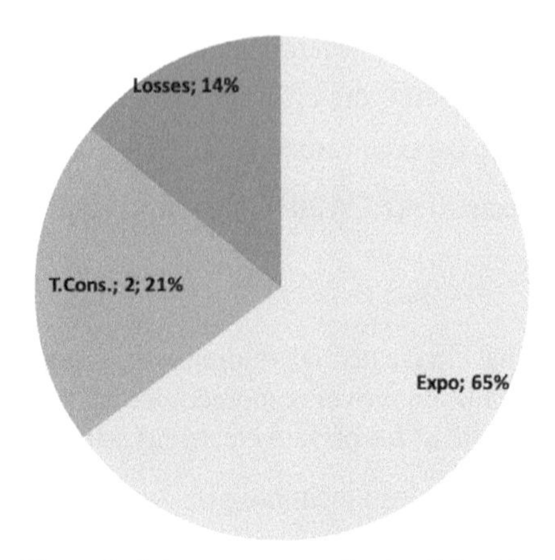

Figure 9. Primary energy production Colombia.

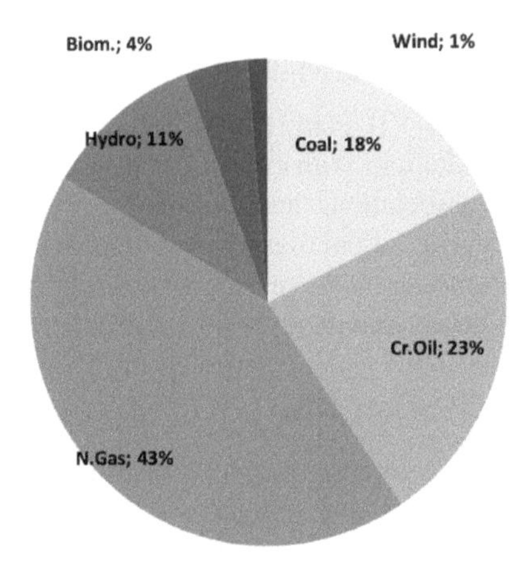

Figure 10. Local primary energy Matrix 2015. Sources: DNP 2017: 12.

connected to the national electric system in Colombia and only two solar grids existed before 2018 (UPME 2020).

To address the potential of renewable resources, the Ministry of Mines and Energy held in October 2019 the first successful auction of new renewable energies in its history, allocating 1,300 MW to 5 wind and 3 photovoltaic projects. They were won by 7 generating companies and 22 regional distributors. Within the plans of the country is to approve another auction shortly, to reach a total of 2,200 MW in 2022. This would mean the use of close to 13% of NREs in the total current electricity generation capacity of the country. The renewable auction closed with an average price of Col$ 95/kWh (US$ 2.75/kWh – Exchange Rate: Col $ 3,450 per 1 USD), which competes favorably with the generation of non-renewable sources.

The incursion of Colombia in the development of NRE has been possible due to the updating of favorable legislation through Law 1715 of 2015, which together with complementary regulations allows investors to:

- Exemption of import duties on equipment;
- VAT exemption for equipment's nationalization;
- Investment discount of income tax through accounting depreciation;
- Introduction of Net-metering mechanism for distributed generation;
- Obligation of the interconnection of new generators to grid networks;
- Participation of the big owners of the current networks in the tariff scheme. For small owners of distributed networks, stability in purchase prices is not yet guaranteed.

Green Industrial Policies for Technology Generation, Exports and Jobs with NRE

Among the countries examined in the present study, two have policies that not only encourage the installation of new renewable capacities. They also promote them with industrial policies, the development of environmental technologies, companies and jobs and also export creation for Germany and China.

Germany, along with the USA, Japan and other European nations, has been one of the pioneers in the creation of NRE: "The Renewable Energy Act (EEG) in Germany is the foundation for the worldwide expansion of renewable energies and its application led to their cost decay, which was anticipated only by very few experts. The basics of the EEG has been copied in more than 50 countries supporting the use of renewable energies" (Waffenschmidt 2017).

The tariff system developed for residential and commercial energy consumers in the country made them pay more for the services. This allowed for subsidizing solar, wind and biomass installations, which were initially expensive and non-competitive in a market economy. This made it possible for many companies to pioneer and focus on the creation of new technologies and industries. Also, it permitted many persons and small producers to install NRE systems without facing financial risks.

In Germany, an environmental and political economy was formed that encouraged the development of the social and business environment. It stimulated the innovation of technologies that were made in harmony with environmental needs, understanding the challenge of global warming as a great opportunity to create companies, employment, income, technology and exports. A clear example of the importance that green business development has had in Germany is that its participation in the world`s GNP was 4.6% in 2016, while in green markets that participation amounted to 14% (BMU).

The global size of 'green markets' yields considerable figures: "The global market volume for environmental technology and resource efficiency ended the year 2016 at 3,214 billion euros. This figure is the sum total of the market volumes for the six green tech lead markets: (1) Energy efficiency: 837 billion, (2) Sustainable water management: 667 billion, (3) Environmentally friendly power generation, storage and distribution: 667 billion, (4) Material efficiency: 521 billion, (5) Sustainable mobility: 421 billion, and (6) Waste management and recycling: 110 billion euros." (BMU 2018: 7).

It is estimated by BMU that these markets will grow globally at an annual rate of 6.9% until 2025 (8.8% in Germany) and that they should continue to do so vigorously in the future in order to reach the goals of the 2015 Paris Environmental Agreement and the UN's 2030 Agenda for Sustainable Development.

The German business net had grown solidly in the environmental sector and 2016 it made up 15% of its GDP (projection of 19% by 2025), employed 1.5 million people (BMU 2018) and had broad participation of companies, from giants like Siemens to many small and medium-sized enterprises. Ninety percent of the companies in the sector sold less than 50 million Euros, 43% were small companies that invoiced less than 1 million Euro per year and three-fourths of the producers employed 50 or fewer workers. Their average export quotas ranged from 29 to 48% in 2016, indicating large participation of companies from all over the country. The broad export projection has been incubated thanks to the sustained support of public policy for environmental entrepreneurial activity.

German wind turbine technology is well-positioned. Siemens-Gamesa is the largest company in the world (104 GW pipeline capacity) and had about 1,400 projects in more than 90 countries in 2019. Together with Vestas from Denmark and General Electric from the USA, they are the three largest working on wind technology in the world. In fourth place is another German company, Enercon (21 GW pipeline capacity), and the sixth-largest wind-company is Senvion, based in Hamburg (18 GW pipeline capacity).

Presence of German companies in the solar markets is also significant; Centrotherm Photovoltaics was already founded in 1976 in Baden-Wuerttenberg, while the best known, Solar World, was created in 1988 and has a presence in the USA and all continents. However, german production of solar panels has been reduced due to strong competition from China and Asia, having then its companies reoriented toward the production of inverters and electronic controls, keeping themselves active in this industry. Enterprises such as SMA Solar Technologies has a presence in more than 20 countries and a wide network of local, small and medium-sized companies that provide inputs, technologies and create well-paying jobs in Germany.

In China, the environmental production matrix has been marked by state goals established in its five-year plans for the electricity sector. These include mandatory fees for the purchase of green energy by large state distributors, which entails the stability of sales for new generators and allocation of negotiable Green Certificates on the stock exchange, which improves their profitability. The feed-in-tariff strategy also encouraged large investments (up to June 2018) and the adoption of distributed systems. Since 2015 (CREO 2017), the government authorized auctions for the largest generation of electricity to promote the fulfillment of the goals of production of 15% of its electricity with renewable resources in 2020 and 20% in 2030.

The "Cleantech Investments Report", which includes renewable energy and the environmental sector, show how they have quickly grown since 2007 (425 million Euro), up to 18.5 billion Euro in 2012. This has been a mix of market-driven growth through great stimuli, and state goals to diversify both its energy matrix and the adoption of green technologies (EUSME 2014). In order to reduce the very high levels of pollution and simultaneously turn green innovation into a reef of local and national development, these achievements (Xielin et al. 2012) have been guided by principles such as:

> "We know that innovation requires a firm market demand. In the environmental area, such demand will not develop or persist without strong regulatory frameworks. Clear standards and enforced regulations created equitably by the government are what create the market for environmental innovation. Without strong, clear, stable, and uniformly-enforced regulations and standards, there is no market and hence no incentive for investment in development and wide deployment of innovative environmental technologies."

The creation of a large national market and export companies meant that 52 Chinese companies were within the 200 largest in the world, according to "The Clean200 Index 2018 Q3". They were followed by 34 from the USA, 19 from Japan and 9 from Germany. Its latest report places Toyota Motors as the first major green company on the globe for its sales of hybrid vehicles and secondly, Siemens for its wind, water and control systems for NRE. However, despite the wide presence of Chinese companies in the top 200, they have recently suffered a decline from previous reports due to the ongoing commercial war with the US. The US has imposed high import tariffs for green technologies imported from China and given the high competitiveness that solar and wind technologies have achieved; the Chinese government has eliminated subsidies for renewables since June 2018 and encouraged auctions as a strategy to generate low-cost electricity toward the future.

With solar technology, whose pioneers were US and German companies, Chinese companies have recently become the biggest in the world. "China is currently the world's largest solar energy market in terms of solar energy production and solar power consumption. IEA suggests that China's rapid increase of 53 GW capacity in 2017 was the major contributor for the global solar energy capacity growth. In early 2018, China surpassed its 2020 solar PV target outlined in its 13th Five-Year Plan. Since 2012, China's share of global PV demand has grown from 10% to more than 55%. Many China's major solar energy companies are nowadays the world's leaders in this market" (The Clean 200 Index, 2018 Q1). Among them, Jinko Solar, based in Shanghai, was ranked as the largest exporter in the world in 2018, sending 11.4 GW of modules per year all around the world. In third place was Trina Solar (Changzhou headquarters), while other large Chinese producers and exporters were J.A. Solar (Shanghai, 27 GW cumulative module exports), LONGI Solar (Xi′an headquarters), which is the world's largest manufacturer of mono-crystalline products and expects to reach 45 GW of solar-wafer production by 2020. Other major Chinese exporters are Telesun (headquarters in Suzhou) and Risen Energy (Ningbo). In the market of inverters and electronic equipment for the solar industry, the Chinese companies Huawei and Sungrow occupied the first two positions in the world in 2018 with the German SMA Solar Technologies being the third.

The largest wind technology company in China is Goldwind (15 GW pipeline capacity). It is the seventh-largest company in its field and specializes in medium capacity turbines (0.6–1.5 MW). The eighth-largest global company is Sinovel (14 GW capacity) which is a leader in big wind sets (5–6 MW) and has an experimental base for turbines of up to 15 MW, the largest in the world.

The extensive network of companies in the NRE sector in China has meant that the employment situation (direct and indirect) in the energy sector is expected to undergo a significant transition in the future. A large part of jobs will be moved from the traditional energy sectors (mainly coal: 5.8 million in 2015), which would be reduced to 1.3 or 2.6 million by 2050. It is expected that this reduction would be replaced by jobs generated in renewable energies. These jobs would grow from 3.3 million in 2015 to between 11.5 or 14 million in 2050, depending on the probable scenarios (CREO 2017). Jobs in solar industries would represent about 60% of the total, following those in the wind sector.

Regarding technological development, a report issued on January 2019 by the Global Commission on the Geopolitics of Energy Transformation, which was set up by the International Renewable Energy Agency (IRENA), points out that "China has taken a lead in renewable energy and is now the world's largest producer, exporter and installer of solar panels, wind turbines, batteries and electric vehicles. China also has a clear lead in terms of the underlying technology, with well over 150,000 renewable energy patents as of 2016, 29% of the global total. The next closest country is the U.S., which had a little over 100,000 patents, with Japan and the E.U. having closer to 75,000 patents each." However, the International Patent Office also notes that the total of new renewable technology patents has tended to decrease in recent years, stating that "the green energy technologies required to curb emissions exist, yet the obstacles to their diffusion are manifold." (WIPO 2018: 21).

To conclude, countries that have had better institutional conditions, business environment and market to develop NREs have been clearly and consistently supported by state policies. These policies were incubated through direct and indirect resources (such as private sector subsidies) and the conditions to pioneer NRE investments and installations. These countries considered that the energy transition was not only an environmental challenge but also a national opportunity that could mean the creation of state and private companies that served local markets and were projected as competitive exporters of new technologies. But even the leading countries have limits to their developments: these can be seen through the moderate diffusion of NREs in order to replace the fossil fuels locally and in the rest of the world. Thus, facing the serious risk of continuing to accumulate CO_2 and intensifying climatic catastrophes globally.

It can be concluded that markets, financing, and technological systems currently prevailing in the leading nations should be reproduced and adapted worldwide. They should be developed at a much higher rate than the status quo since the current implementation is far from offering guarantees of achieving the necessary energy transition of the planet before serious damage occurs to ecosystems and future generations. New political and social dynamics that reinforce faster implementation are required to ensure energy transformation throughout the globe. The European Parliament has just declared the "climate emergency" (November 2019) to affirm and accelerate the transition. The whole world should join in these actions with much greater decisiveness and urgency.

References

BID – SENECO. 2019. Potencial de Energías Renovables de la Ciudad de México.

BMU – Bundesministerium fuer Umwelt. 2018. Green Tech made in Germany – Environmental Atlas Technology for Germany. Berlin.

BP – British Petroleum Company, Statistical Report, 2018.

CEPF – Centro de Estudios de las Finanzas Públicas. 2017. Cámara de Diputados del Gobierno Federal, México DF.

CEPFP – Centro de Estudios para las Finanzas Públicas. 2017. Incentivos a las energías renovables: una alternativa de ahorro para el Presupuesto del Gobierno Federal – Cámara de Diputados. Ciudad de México.

Chiu, D. 2018. The East is Green: China`s Global Leadership in Renewable Energy. CSIS.

CREO – China Renewable Energy Outlook. 2017. Energy Research Institute of Academy of Macroeconomic Research – NDRC. China National Renewable Energy Center.

DNP – Departamento Nacional de Planeación. 2017. Energy Demand Situation in Colombia. Bogotá.

EPRS – European Parliamentary Research Service. 2018. India: Energy Issues.

EU SME – European Union Small and Medium Enterprise Centre. 2014. The Green Tech Market in China.

Fraunhofer ISE. 2018. Recent Facts about Photovoltaics in Germany. https://www.pv-fakten.de.

FS – Frankfurt School & UNEP Centre/BNEF, 2019. Global Trends in Renewable Energy Investment.

García-Alvarez, M. T. and Mariz-Pérez, R. M. 2012. Analysis of the success of feed-in tariff for renewable energy promotion mechanism in the EU: lessons from Germany and Spain. International Congress on Interdisciplinary Business and Social Science. University of La Coruña, La Coruña 15071, Spain.

González-Jiménez, E. 2016. Costa Rica 100% renovable: Claves y lecciones de una política eléctrica exitosa. Friedrich Ebert Stiftung.

IEA – International Energy Agency. 2019a. CO_2 emissions from fuel combustion. Statistics.

IEA – International Energy Agency. 2019b. World Economic Outlook.

IEA. 2019c. Country Policies. https://www.iea.org/policiesandmeasures/pams/costarica/name-160892-en.php?

IRENA. 2019. Renewable Power Generation Costs in 2018, International Renewable Energy Agency, Abu Dhabi.

MINAE – Ministerio de Ambiente y Energía de Costa Rica. 2015. VII Plan Nacional de Energía. San José, 2015.

OCDE. 2019. Energy Statistics. Paris.

Segura-Elizondo, O. 2015. Ministry of Environment and Energy. Distributed Energy Policy in Costa Rica.

SENER – Secretaría Nacional de Energías Renovables. Junio 2018. Reporte de Avance de Energías Limpias. Reports 2017 and June 2018.

The Clean 200. 2018 Q1. Carbon Clean 200™: Investing in a Clean Energy Future. https://www.asyousow.org/report/clean200-2018-q1.

UPME. 2020. Boletín estadístico de Minas y Energía. https://www1.upme.gov.co/InformacionCifras/Paginas/Boletin-estadistico-de-ME.aspx.

Waffenschmidt, E. 2017. The Renewable Energy Act in Germany: its basic idea and recent developments. Technische Hochschule Koeln, Conference Paper.

Wassermann, S., Hauser, W., Klann, U., Nienhaus, K., Reeg, M., Rhiel, B., Roloff, N. and Weimer-Jehle, W. 2012. Renewable Energy Policies in Germany: Analysis of Actors and New Business Models as a Reaction to the Redesign and Adjustment of Policy Instruments. CIRIUS – University of Stuttgart. Stuttgart.

World Bank Group. 2020. Global Solar Atlas. https://globalsolaratlas.info/map?c=11.695273,8.173828,3.

Xielin, L., Strangway, D. and Zhijung, F. 2012. Environmental Innovation in China. WIT Press, Southampton, U.K.

Xin-gang, Z., Yu-zhuo, Z. and Yna-bin, L. 2018. The Evolution of Renewable Energy Price Policies Based on Improved Bass Model: A System Dynamics (SD) Analysis. MDPI.

Zárate, D. and Ramírez, R. 2016. Matriz Energética de Costa Rica. Friedrich Ebert Stiftung. San José.

CHAPTER 10

The Sea and the Hegemonic Condition of Economic Power

Challenges for Colombia

Enrique Alfonso Ochoa,[1,*] *Madelcy Del Carmen Pedroza*[2] and *Jaime Eduardo Gonzalez*[3]

Introduction

The historical evolution of humanity has shown that cultures develop in better conditions when they are allied to the development of international trade routes and therefore of the oceans. That is why countries that have administered power in a hegemonic way have simultaneously had a relationship with the development and control of the ocean. At present, there is a factor that is associated and complements the previous variables, which refers to satellite technology, that in essence has allowed the control of international trade relations to be developed with higher levels of efficiency, effectiveness and efficacy (Fairbanks and Lindsay 1997, Mahan 2011).

This chapter deals with criteria that allow us to understand the relationship between economic development, maritime transport and the condition of power of the countries that use the oceans efficiently and effectively and their impact on the development of humanity with regard to being in present times, by using satellite systems, a process of essential use in the transport of cargo and more specifically in the interoceanic domination and with this expansionism.

Colombia's situation is carefully reviewed with respect to the use of its maritime platform and what is being done with it today since the largest poverty settlements

[1] Faculty of Economic, Administrative and Accounting Sciences, Universidad del Sinu Elias Bechara, Zainum, Av. El Bosque Transv. 54 No. 30-453, 130001, Cartagena, Colombia.

[2] School of Accounting Sciences, Remington University Corporation, Calle 51 # 51 27, Parque Berrío, 050001, Medellin, Colombia.

[3] A Faculty of Economic, Administrative and Accounting Sciences, Comfenalco Technological University Foundation, Campus A Barrio España Cr 44 D N° 30A – 91, 130001, Cartagena, Colombia.
Emails: madelcy.pedroza@uniremington.edu.co; gonzalezdj@tecnocomfenalco.edu.co

* Corresponding author: enrique.ochoa@unisinu.edu.co

in that country are on the seashore, e.g., the situation of the Province of Choco and the province of La Guajira, where the presence of the state is a socio-economic and geopolitical lacking and the central government is in debt with these regions. In addition, a diagnosis is made on the incidence of the socio-economic evolution of advanced civilizations and their relationship with the implementation of maritime policies. Furthermore, to achieve the development of this topic, it was necessary to address precise data of universal history, the discovery of America and other relevant facts in the course of the consolidation of civilizations and their relationship with the efficient and effective use of the oceans (Rivarola 2011).

The Ocean as the Main Means of Communication and Economic Development of the First Civilizations

It is convenient to consider that, in the transition between prehistory and history, the greatest finding to highlight the appearance of writing as a means and definitive way to establish and strengthen communication between men and/or civilizations, even though there was already a connector between the peoples through trade and at increasing scales: the oceans.

These events took place in the ancient Egyptian civilization with the help of the Nile River, which brought the different minerals from the lakes 'Alberto and Victoria' of Central Africa, where this majestic river was born. In its generous journey, it enriches the soil in its path, not to mention the magnificent landscape sights of the seven waterfalls, which allow the connection of freshwater basins between the continental shelf and the oceans; all-around wonderful conditions where the environment, its conservation and preservation, provide man with natural conditions of well-being and quality of life (Garrison et al. 2003).

However, as provided by Divine Providence, Egypt made the great mistake of transferring the ocean trading to 'the seafaring' Phoenicians, who fully understood the opportunity they had been given and of course, did not miss it (see Figure 1).

The Phoenicians assumed a clear and forceful role in the management of trade between "Mesopotamia, the Tigris and Euphrates Rivers and the Mediterranean" now known as the Persian Gulf (see Figure 2), inhabited by Iraq, Iran, Kuwait, Saudi Arabia, Bahrain, the United Arab Emirates and Qatar (Aubet 2008).

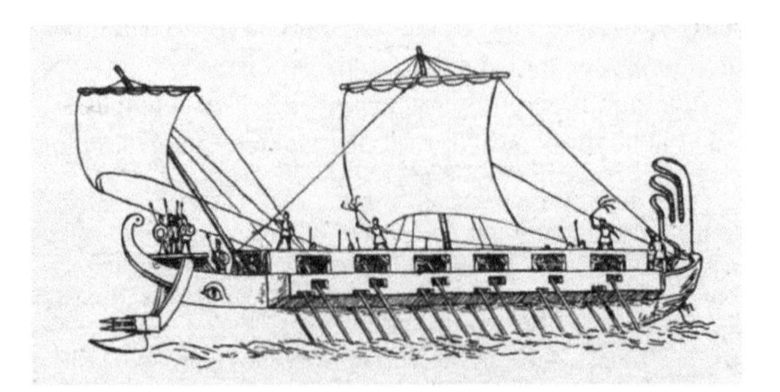

Figure 1. Model Phoenician boats in the first millennium B.C. Source: title: definition of Phoenician culture, author: Javier Navarro, July 2017, www.definicionabc.com/historia/cultura-fenicia.php.

Figure 2. The Persian Gulf. Source: Infobae (2019). Recovered from: https://www.infobae.com/america/eeuu/2019/10/03/los-regimenes-de-rusia-china-e-iran-preparan-maniobras-navales-conjuntas-en-el-golfo-persa/.

The 'seafaring' of the Phoenicians was extended in such a way that this entrepreneurial and mercantile civilization enjoyed a good moment in the history of civilization. However, we all know that everything that begins has an end. It was then when an eminently seafaring culture appeared. It found enough reasons and spaces to give a strong and definite blow to the Phoenicians, who roamed and traded through the Mediterranean, more as intruders than as masters of the territory; 'The Cretans', to whom the story refers to, acted as leaders in the 'Iron Age', thanks to the large deposits of this metal that existed in their zone. The Cretans also had no additional option but to be sailors. Remember that the Island of Crete is located in the center of the Mediterranean and its geopolitical position was well used. They could control the entrance to the Mediterranean both tactically and strategically, especially in the area from the 'Suez Canal'.

In the midst of the processes, and more specifically between the Medians, as the Persians were known to the Greeks, and the Greeks, there were the events known in history, such as the Median, or Greek-Persian Wars, whose main objective was the control and dominance of the trade routes in the Mediterranean, a situation that occurred between the years 490 and 478 B.C. In this scenario, we can see again that the sea as the background of changes and transformations in the history of mankind. It is important to consider a principle that has been put into practice in the development of civilization and that still remains an important element to keep in mind for all cultures that intend to undertake actions that allow them to expand their territory. It is that the dominant powers assume control over the dominated ones (Mastanduno 1997) and usually, this is accomplished through the imposition of the military figure, the one that in times of war allies itself to strategic locations, therefore, it must be interpreted, that definitely the great changes that have taken place throughout history, have taken place at sea, an indispensable ally (Mirzoyan 2010).

The situations presented by control and hegemony at sea reveal a permanent challenge for the control of trade in the Mediterranean. This is how an awakening of European peoples presents itself and that is when the Greeks prepare to assume the command and control of the situation since they were the most advanced civilization on the European coasts. This contributed to the development of its large cities 'Sparta and Athens'; the first, a civilization entrenched in the culture of war, with a clear and forceful objective: 'Control of the sea'. On the other hand 'Athens', directed efforts oriented towards the intellect as a fundamental element to project growth and development and refers to the land of 'Socrates, Plato and Aristotle' among others, who are important referents when talking about science.

The truth is that even today, the current scenario for the domination of civilizations is 'the sea', and it was the wars at the time that allowed Greece to take power on behalf of the European continent. Everything was going well and the Greeks managed to establish the domain of trade relations in the Mediterranean; there was so much euphoria that around the Spartans, the physical culture, took a lot of strength and representativeness to such an extent that the soldiers became narcissistic losing their identity, respect and full awareness around their military culture.

Due to the above circumstances, an advanced process around the control of the Mediterranean is undertaken by Rome. On this occasion, the great entrepreneurs were the Romans under the leadership and direction of one of the most important and representative men of ancient history, Julius Caesar. Acting under the principles, navigation in the service of civilization and all roads lead to Rome (Eubank et al. 2019), he opened a space for participation entrenched in the geopolitical culture, seeking connectivity of the territories and the continental platform through the optimal use of the sea and the consolidation of roads. In such a way, that the expansion toward the north of Europe was achieved to expand and reach into the rest of the old continent. This process was known as the 'Roman Empire'. Without neglecting that the expansion into Africa, which should have been first by sea, finding complex situations that should have been overcome, such as the confrontation against Carthage. One of the fiercest naval wars that after all was controlled by the Romans; events that occurred between the years 264 and 146 B.C.; all for the control of the economy and the trade routes in the Mediterranean. This was known in history as the Punic wars; it was perceived then that Julius Caesar, possessed in his prospective thinking, the elements that today identify the sustainability of a civilization.

With respect to the empire, there is sufficient evidence in the historical documents that there is a manifest of the abuse of power. The application of the principle of submission and even humiliation, registering clumsy and excessive actions exercised by some emperors such as Caligula, Augustus and Nero among others, leading to discredit the civilization that ultimately dies under the force and power of the invaders of the north, a fact that was known as the invasion of the barbarians.

In the consecutive history of humanity, it is important to highlight that in the bosom of the Roman Empire and the lands of Galilee in a manger Jesus, the son of God, was born (Fiensy and Hawkins 2013). This situation from the biblical perspective, explains the inordinate and barbaric acts by Roman soldiers under the governance of Pontius Pilate. All of this happened because around the pilgrimages and proliferation of the Gospel, Jesus rejected the action of the empire regarding the civilization of the moment. This situation reaches its extreme point with the Persecution, torture, sacrifice, death and burial of Jesus (Fuhrmann 2012). The aforementioned, and in contrast to what happened in the resurrection process, led civilization to fear, secrecy and mystery. Because men were stunned, with the situation of the beginning of the Middle Ages, also known as the millennium of darkness, or 'dark ages', where the backwardness of civilization was imminent because of perplexity, fear and guilt and disrespect for the matters of God and even the Holy Scriptures.

After the processes around the middle ages in the fifteenth century, an awakening appears in medieval civilization, a phenomenon known as the 'renaissance', where characters such as Leonardo Da Vinci, Rafael and Galileo Galilei (Oosterhoff 2012), among others, project ideas and execute actions that contribute with the awakening and eye-opening of the population. It is precisely the end of the medieval era which is characterized by the revival of the seafaring trades, which impulses the discovery of America. As a result of a mapping error made by the Genovese sailor, Christopher Columbus' navigators' team set sail, wanting to reinitiate trading with India. Instead, they ended on the Caribbean coast and later on, they discovered that it was a new continent because in a deeper exploration exercise Amerigo Vespucci made a coastal sailing near the seashore, reaching the same point where he had started from after several years of adventure. Then he fully understood that it was a new land and thus was called America in homage to the daring cartographer, who could demonstrate that the new lands had not been explored by European civilizations.

Once more it was necessary to project resources and processes for the exploration and exploitation of sea riches. In an effort to achieve and exploit new routes and after many attempts, an event that changed the world. America was discovered on October 12, 1492, thanks to the boldness and tenacity of the sailor Christopher Columbus, who set sail from the port of Palos in Moguer and who due to a cartography accident arrived at an island in the Bahamas called Guanahani. Again, the sea is the stage and leading character of one of the most important and relevant events in the history of humanity.

As expected, behind the Spaniards, new cultures of European navigators represented by Portuguese, English, French and Dutch appeared. They were eager to explore new lands and when they discover that those regions that were once known as the Indies, gold was used even to cover the nether parts of the natives (García 1994). This generates a direction toward the exploration and exploitation of the Indies, an event that occurred simultaneously with the discovery of North America (1497–1498) and the conquest of Central and South America, terms used to differentiate the level of the relationship established by the English, French in the north and Portuguese and Spanish in Central and South America.

There were many situations that appeared during the sixteenth, seventeenth and eighteenth centuries, characterized in the south by barbarism, exterminations, the violation of human rights, to the point of being recognized by the critics of history, sociology, law among other disciplines, such as three centuries of barbarism.

The wars in the Atlantic were cruel and ruthless. Vessels were persecuted and even sunk with cruelty sacrificing whole crews of sailors, just for the greed of precious metals, which were supposed to be aboard ships heading to the old continent; from this process in which the results began to glimpse in the 1800s, a great beneficiary appears 'the United States of America'.

After the independence of the peoples of North America, thanks to the unrestricted and representative sponsorship of the French navy, the people of South America achieved their independence. It is important to remember the raison d'être of Simón Bolívar's trips to Jamaica on behalf of Gran Colombia and the way in which this process contributed to the freedom among others of our beloved and beloved Nation, Colombia.

On the other hand, the United States played a role of superior characteristics, being the one in charge of identifying and assuming control over points in the globe that in the first place had a direct relationship with the sea and later on would become bridges and mandatory steps for the control of international trade routes. These points are known as geostrategic points of maritime control; which are: the Suez Canal (see Figures 4 and 5), the English Channel and the Panama Canal (see Figure 3) in addition to the Straits of Gibraltar, Behring and Magellan; and as distribution points, the Canary Islands in the Atlantic Ocean and the Azores in the Pacific (Spivey 2015).

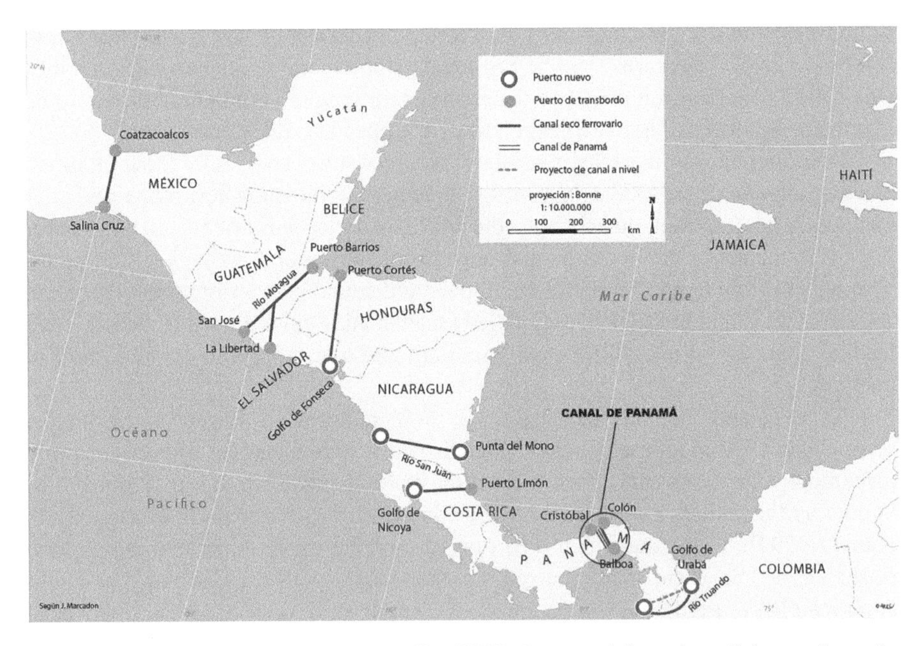

Figure 3. Panama canal. Source: Atlas Caribe (2019). Recovered from: https://atlas-caraibe.certic.unicaen.fr/es/page-67.html.

Other War Conflicts Over Control and Domination of the Sea (Russian-Japanese War)

At the dawn of the twentieth century, the interest of some specific areas of the planet was revealed, and the first to show their military infrastructure were the Russians and Japanese on the scene of the Russian-Japanese War (see Figure 6), where the region

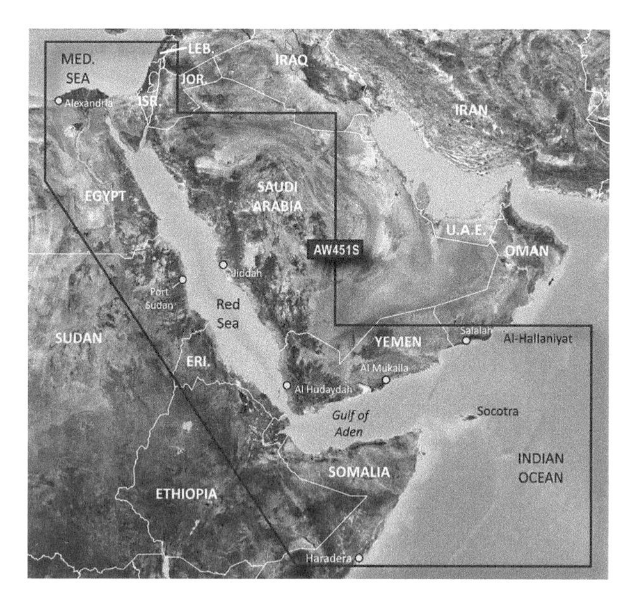

Figure 4. Suez canal. Source: Pinterest (2019). Recovered from: https://co.pinterest.com/pin/413346072056122559/.

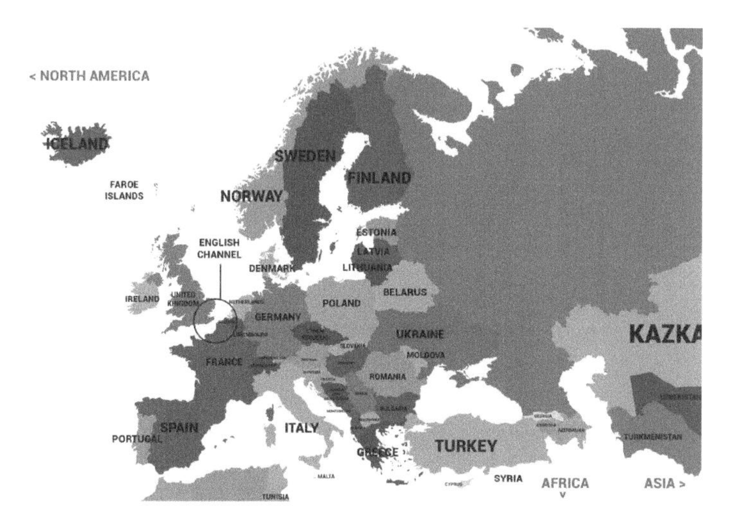

Figure 5. The English channel. Source: Recovered from Pinterest, 201? https://co.pinterest.com/pin/862931978575430327/visual-search/?cropSource=6&h=267&w=300&x=10&y=10.

of Manchuria (see Figure 7), in the southwestern part of China is located. This conflict that was eminently naval and highlighted Japanese power and technology with the ability to make communications interceptions, which contributed determinately to the humiliation that the Japanese gave to the Tsarist troops. After that war, many countries fully interpret the importance of the sea because of the results of that conflict and the onslaught of Japan over Russia. There was an uprising of the Russian people

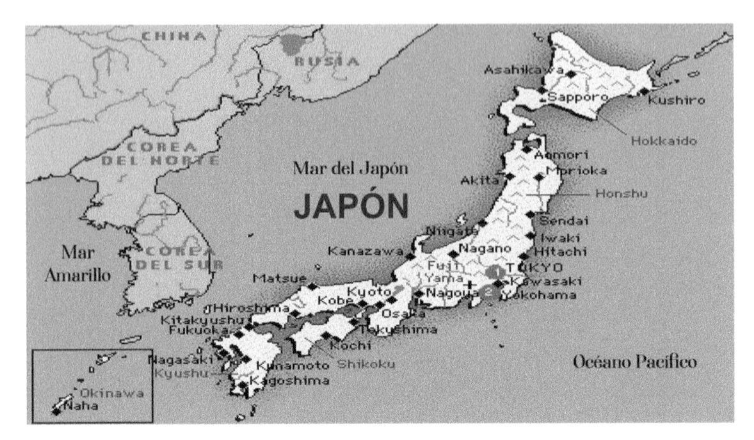

Figure 6. Stage of the Russo-Japanese War. Source: Adapted from Story anecdotes (Luisma and Omayra 2011). http://luismayomayra.blogspot.com/2011/02/batalla-de-japon-contra-rusia-1904-1905.html.

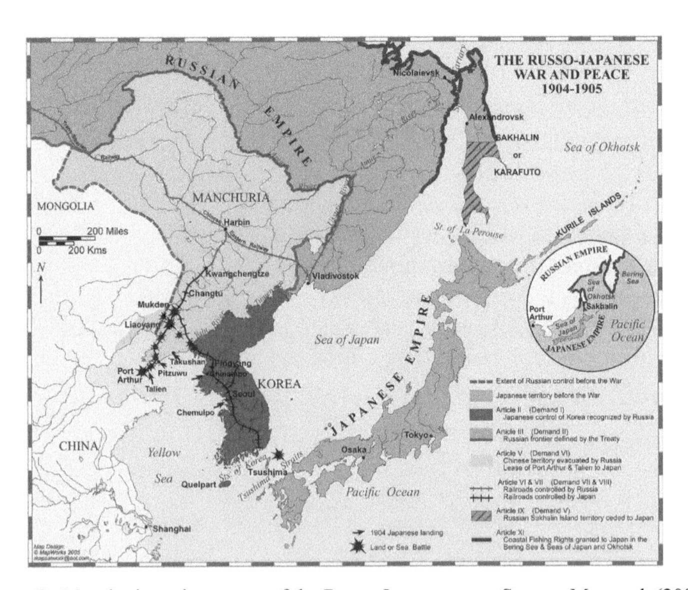

Figure 7. Manchuria region, scene of the Russo-Japanese war. Source: Mapwork (2005).

that generated the Bolshevik revolution: cause, effect and consequence of the birth of Communism. It was ideologically entrenched under the statements of Karl Marx, in his document called *Das Kapital*; and the leadership of the process was under the autonomy and control of Vladimir Ilich Ulyanov, also known as Lenin. Again the sea was at the center of an event of great importance in the history of mankind (White 2015).

Already in the twentieth century, the First World War (1914–1918) and later the Second (1939–1945) showed the importance of the oceans the sea as a crucial element for the control of the planet and, of course, production, trade and with them, politics. Society and the economy, the control and dominion of the sea was not only an argument for international war conflicts but also that around the oceans the future of humanity was defined in confrontations, which had a no different explanation in their interpretation on who controls the sea controls the routes of international trade and, therefore, controls politics and the economy (Ahmed 2015).

Regardless of any reason or argument made about the causes of the war conflicts of the twentieth century, in the case of the First and Second World Wars, having sea-coast access to the ocean for the Germans, was reason enough to involve in the conflict all the powers of the world of the time in a war.

That is why, the Triple Hegemonic Alliance of power was still maintained, the epicenter being the United States, and backed by Great Britain and France. The situation that was evidenced in-depth in the events of the First and Second World War, where the Germans on both occasions lashed out their warring strategy on the English Channel. They knew the importance of its strategic position, and it was considered the only exit to the international trade routes of the countries that are located in northern Europe; only with an aggravating factor of crucial importance, both times the United States came out in defense of its great allies. It is important to remember that the Statue of Liberty was installed in the Bay of Manhattan by France, as a symbol of alliance and emotional ties with the country of Uncle Sam.

If the events and consequences of the conflict actors are appreciated, it can be seen that the G-8 or countries that produce 85% of the world's GDP on average, have a common factor that today they demonstrate through innovation, technological development and represented in the generation of added values and advanced production typified as industrial. They are the same eight that were involved in the two wars simultaneously, which were developed and decided ruling over the oceans and, the great reason, 'the control of the sea'.

It is also possible to identify that the most benefited from these conflicts was the United States of America, on whose continental platform there is no skirmish. The benefits were of such magnitude that the post-war wars, at the end of the 1940s, this country owned 52% of the world's GDP and through the Marshall Plan organized what is now known as western capitalism by consolidating a strategic alliance of great respect in Asia; Japan was the same country that received two atomic bombs in the development of the Second World War.

Therefore, it is said that whoever owns the sea and does not put it to good use is punished in his condition of inefficiency and ineffectiveness for the administration of natural resources and geostrategic positions.

Alfred Mahan, in his geopolitical thinking, states that a leader is one who formulates the vision and is the one who points the way, sets the guidelines and explains the requirements. This was interpreted and put into practice by the US naval force and to which the great advances of the late nineteenth and early twentieth centuries are due with the purpose of being the society of the superpower country, a culture that implies an expansion in its needs and this is vital and essential for the evolutionary development of the nation (Bridge 2001).

The Sea as the Leverage of the USA Economy (E.G.)

There are some elements of great importance, which highlight the power and hegemony that the United States has on the planet, and it is directly associated with its capacity of expansionism, which has battled and defended it tooth and nail in different scenarios; some under the direct intervention in land negotiations, influenced by two characteristic elements:

- The state of weakness of the people or the culture that sells them, and the dimensions of the association with maritime borders.
- The successful geo-strategy and political decision to implement the 'Marshall Plan' in the post-war scenarios, given the situation of strength that Uncle Sam's country had at the end of WW2 and everything related to the Cold War proselytism. That nation undertook in competition with its strategic ally in the scenarios of the Second World War, 'the Union of Social and Soviet Republics' or USSR.

The following are the main scenarios, where the United States of America, as an example of expansionist leadership, carried out its purchase maneuvers and/or acquisition of territories:

- Eighteenth century: The events presented in the Treaty of Paris (1783), where Great Britain contributes to the definition of American borders, assigning it to the US, Machias Seal Island and North Rock. This situation seems an unclosed chapter with its bordering neighbor, Canada; furthermore, the dispute of the Indian Stream territory, a situation that was resolved by the Webster-Ashburton treaty (1842).
- Nineteenth century: The purchase of the Louisiana region at the beginning of the 1800s with the intervention of Robert Livingston, being the US President, Thomas Jefferson; territory for which the sum of 15,000,000 million dollars was paid to France. In the referenced negotiation, a relatively small part was ceded from Spain in 1819 in the same negotiation that was made by the territories of Florida (Explanation of the cause, reason and prevailing existence of the Spanish language in this emblematic region of the United States, where even culture has a representative bias toward Hispanic culture).
- After independence, from the territory that today comprises West Florida, which was owned by Spain, it was declared by President James Madison through possession of the United States in 1810, the United States assumes and assigns it to its continental platform.
- At the end of the second decade of the 1800s and through the Adams Onís treaty, the Spanish royal crown ceded part of eastern Florida to the United States of America. It included a representative sector of the Central Colorado area; negotiations were all conducted under the imposition of power and the hierarchy expressed by the United States in the face of its weakened adversaries.
- At the beginning of the second decade of the 1840s, a representative area between the USA and Canada, which once served as a Canadian colony and through the Webster-Ashburton Treaty, the referenced territory was awarded to the USA.
- Even in present times, and even in modern civilization and of the present generation, there is uneasiness over the way the United States acquired the

region of Texas, following a determined diplomatic pressure exerted by the Prime Minister of the States of Mexico. Diplomat Joel Roberts and his work team intervened within the process supported by the military forces, who pressed the representative of the Mexican government Antonio López de Santa Anna, where there were military clashes in the so-called battle of San Jacinto. This gave the final event where Mexico recognized losing Texas, and New Mexico in the Treaty of Guadalupe Hidalgo, at the end of the 1840s. There are still feelings of disagreement about the situation presented.

- Another representative part of the Mexican territory, which ultimately remained in the hands of the United States of America, in the framework of the Mexican-American War was presented between 1846–1848. It ultimately generated payment for 15 million dollars, with a transfer of 3 million from the Mexicans to the United States; a situation that is of permanent uneasiness in the present generations of Mexican society.
- Twentieth century: Alaska became American territory after it was negotiated with Russia in 1867, and Hawaii annexed to the US territory in the 1900s.
- The reason, cause and consequence of being a Cuban territory of the United States (Guantanamo) are due to the fact that this area of the communist country was under US protection through the Treaty of Paris at the dawn of the 1800s.
- Philippines: A territory that was also acquired through the Treaty of Paris in 1898 and achieved independence in 1946 through the Treaty of Manila.
- There are other North American territories of lesser magnitude but all have a common factor: a relative incidence with the maritime condition; these are Guam (Western Pacific), remote minor islands, American Samoa (west of the Cook Islands and north of Tonga Island), Northern Mariana Islands (between Hawaii and the Philippines), US Virgin Islands (Caribbean Sea, part of the Virgin Islands archipelago).

When reviewing each of these previous possessions of the United States, it can be agreed that they are all associated with geopolitical and strategic points, which must surely be used in any circumstance of urgency, where the superpower must intervene in order to defend its interests.

Additionally, an anonymous thought is observed and it states that Globalization is an advanced phase of the United States of America for the control and dominance of planet earth (Tsegay 2016, Wang et al. 2018). This thought that contrasts with that of the current president of that nation, Mr. Donald Trump, who hinted that the opening processes at most in his government are not shared, and he even has expressed that America for Americans (Scarfi 2016). President Trump has also stated that if a business organization in his country needs personnel to work, the US citizens should be preferred over foreign citizens of the same territory, interpreting this that the US took a turn toward economic protectionism in the times of the successful businessman.

Notwithstanding the aforementioned, it is of the utmost importance to interpret that the Marshall plan was nothing more than an expansion plan of world capitalism. Despite having been carried out in the postwar period, it was strategic, methodological and carefully designed to ensure capitalist growth and strengthening to the point of having put on its side, the nations that were part of the axis in the Second World War (Germany, Italy and Japan) process that is still perceived today with extreme strength.

When considering the phenomenon presented by globalization, assumed through its engine, economic openness, it was fully appreciated that it was not precisely expansionism in terms of the acquisition of new territories. It was a situation in which politics of growth and penetration in other regions. It had a somewhat different intention but in the same way of increasing the ideals and cultures of the people aggressive and strategic mercantilism, which was perceived, after all, a different way of implementing geopolitics, and in fact, the great beneficiaries were the powerful and developed countries, while the third world, although it is true that they achieved an excessive invasion of products in their territory that really generated dynamics and acceleration of economic factors. It was also true that there was abused in their good faith of being consumers, who in turn supported the globalization process with relatively low conditions of exploitation for their interests.

Another principle that has prevailed in the mentality of the winners The United States of America is that "it is more difficult to sustain than to achieve first place". This has been achieved by the US power through policies, sufficiently planned and based on the conservation and preservation of natural resources as a condition that shows sustainability and sustainability in absolutely everything they do. The situation that is fully demonstrated in the implementation of quality standards based initially on the standards "international standard organization", for its acronym in English, ISO, and of which there is one for each of the clients and allies of business organizations. Among others, the consumer client, the internal client or member of the organization, the medium environment, industrial safety and corporate social responsibility as essential conservation and preservation factors (Stokes and Waterman 2017).

The Sea as a Leverage of China's Economy

Outside the scenario, the incidence of the predominant power in the late nineteenth century, during the twentieth century and in the early twenty-first century, from a different perspective and point of view, it is feasible to consider, that through theories Friedrich Ratzel (1844–1904):

> "Whoever assumes a geopolitical interpretation, the extension of the territories would become an essential element for the purpose of gaining participation spaces, while considering that the combination between the land platform and the use of seas, would become a combination of high representativeness for the development and evolution of civilizations. These elements that were published in his book "Being and becoming of the organic world".

In addition, if the military thought of Napoleon Bonaparte is interpreted, who in the fullness of his principles and doctrines, considered that it was necessary to expand its hegemony toward the control of the oceans and that was precisely trying to address them where its fragile or weakened point was known since at the time he did not fully understand that maritime wars are totally different from terrestrial wars due to the contrary scenarios.

Napoleon Bonaparte expressed When China wakes up the world will tremble (Andrade 2017). It is evident that China has woken up. It seems then that the expression of Chinese Prime Minister Deng Xiaoping (1904–1997) in his famous phrase It doesn't matter if the cat is white or black, the important thing is that it chases mice (Xiang 2014). Which arose from the visit he made in the year 1960 to its Asian neighbors; South Korea and Singapore, which were developing in an accelerated and forceful way, was when this famous phrase was pronounced to their comrades, advisers of the different fronts that the direction of a country has to assume.

In present times the expansionism of China, which with regard to the commercial war with the United States of America, the Asian giant was able to enlist its artillery for control and international trade routes, to the point of having five of its most important ports among the top ten most important ports in the world; number one is Shanghai Port, number third is Shenzhen port, number fourth is Ningbo-Zhoushan Port, number fifth is Guangzhou and in number eight is Qingdao port.

Economic Analysis of the Countries Involved in the Study (the USA, China, England, France, Germany, Italy, Japan, Canada, Russia (Union of Social and Soviet Republics, USSR) in Relation to the Greater Volume of Movement of TEUs in the World (2019)

It is looming after making a judicious analysis of the countries that continue to mark hegemonically statistical data on the economic activity today, among which power is disputed by two giants, the US and China, and it is perceived that the movement of TEUs of these countries is consistent with the maritime and port structure they have, fully confirming the hypothesis that the economy is hegemonically entrenched at sea and in particular the rivalry trade seems to be overcoming the patience of the rulers of these powers, a situation that today, according to the interpretation of the international geopolitical panorama, there is an alleged chemical war aimed at directly affecting the economy as a way of measuring the pulse and its attack on the two giants the USA and China (see Table 1). It is important to mention that it is precisely the European powers that are most affected in the development of the alleged chemical warfare (Hoffmann and Kumar 2010, Lloyd and Naisbitt 1994, Ventrella 2012).

Based on the foregoing and as can be seen in Figure 8, China has strengthened its hegemony over the sea, which we have summarized in three key points:

- The TEU movement at the international level is led by China, followed by the United States, the European powers, Japan and Russia.
- Regarding GDP, the United States leads the process followed by China Again, Japan and the European powers.
- The hypothesis that the economic power of the countries in the history of humanity, be they, cultures, civilizations, empires, that leads the economic power, its essential ally of the head is the sea and with this, the maritime and port infrastructure is verified and therefore international trade.

Table 1. Nominal GDP of the countries under study. Source: Database Macro Data. Recovered from https://datosmacro.expansion.com/pib.

Comparative: annual GDP			
Countries	**Date**	**Annual GDP**	**Var. GDP (%)**
United States	2019	19,139,884 M,€	2.30%
China	2019	14,380,456 M,€	6.60%
Japan	2019	5,154,475 M,€	0.70%
Germany	2019	3,435,990 M,€	0.60%
United kingdom	2019	2,523,314 M,€	1.40%
France	2019	2,418,997 M,€	1.30%
Italy	2019	1,787,664 M,€	0.30%
Russian Federation	2018	1,658,938 M,€	2.30%
Canadá	2019	1,550,895 M,€	1.60%
Spain	2019	1,244,757 M,€	2.00%
Colombia	2019	327,895 M,€	2.50%

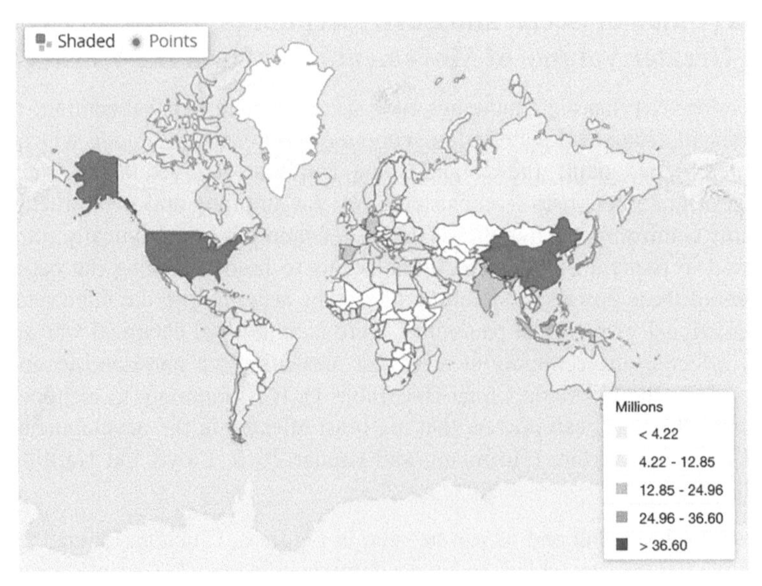

Figure 8. Marine container traffic (TEU: units equivalent to 20 feet). Source: World Bank Database. Recovered from: https://datos.bancomundial.org/indicador/is.shp.good.tu?end=2018&start=2018&view=map.

From the authors' knowledge, it is deduced that the contrast comparison between the volume of TEUs that makes the difference between the United States and China and the relationship between the GDP of each country is explained because it is the North American power producer of technology that of manufactures, since the latter produce the highest volume in quantities but not in economic value (see Table 2).

Table 2. Shipping container traffic (TEU: units equivalent to 20 feet). Source: World Bank Database. Recovered from: https://datos.bancomundial.org/indicador/IS.SHP.GOOD.TU?end=2018&start=2000&view=chart.

Country	Most Recent Year	Most Recent Value
China	2018	225,828,900
United States	2018	54,688,353
Japan	2018	22,433,824
Germany	2018	19,597,633
Spain	2018	17,189,759
United kingdom	2018	11,695,222
Italy	2018	10,547,112
Canada	2018	6,663,690
Russian Federation	2018	6,335,300
France	2018	6,369,200
Colombia	2018	4,125,200

Colombia, Its Seas and Its Privileged Geostrategic Location

The generosity of the divine providence gave the territory of what today is the scene of action and coexistence of the Colombian State. A whole series of virtues, when it is considered that it has diverse qualities under the geopolitical and geostrategic perspective; continental header, connectivity bridge between the two continental extremes through 'Central America' and all climates and topographies that translate into infinite biodiversity, plus the privileges of being part of the equator line.

In the scenarios of history, it was the Caribbean Sea, which bathes the north of our Republic, the witness of the discovery of our continent. Until 1492, it was unknown before the most advanced civilizations of the time. Afterward, our sea and by the performances by Colombian sailor José Prudencio Padilla was considered the sea of independence and later, the sea of freedom. At present times the sea of oblivion, a situation that is due to the waste that has arisen in the two long centuries of freedom. Even being interpreted today, Colombia is a country that has turned its back to the sea without ignoring that it was an expedited scenario for the looting that the European invaders undertook through persecution, humiliation and killing of natives. This generated a catastrophe in what we now call human rights and of course the imposition; in the case for the Central and South Americans of cultures that left much to be desired but due to circumstances of destiny is what marks the idiosyncrasy of our civilization.

As proof of the above with a speech by Evo Morales where he responds to the Europeans, about the alleged debt that Latin America has with the old continent. It is left to see then that the Colombian Caribbean was the expedited means for the implementation of all the modalities of barbarism that occurred with the Spanish invaders. During the years 1492 and 1810, they made our natural heritage a theater of rivalry operations, looting and extermination of cultures with the participation of

Spaniards, Portuguese, English, French and Dutch, and during three great centuries of rivalry, there was ultimately a great beneficiary, the United States of America.

In times of freedom, it is deeply striking that maritime development became an element against the interests of Gran Colombia, political sectarianism mainly between liberals and conservatives were merciless and ruthless as well as clumsy and insensitive to that point. In the events presented by the interest that awoke in the area of the Panama Canal to North Americans; Colombia had no maritime fleet with the capacity to defend the elements of sovereignty. Colombia only had a few wooden vessels that sailed from the city of Cartagena to try to defend the interests of our nation in the canal area.

But the North American potential that had already made its presence in the Panama isthmus had to help and protect the Colombian sailors because the condition of weakness was revealed. Only in the arrival to the canal area, in this regard, the American admiral Alfred Thayer Mahan (1840–1914) in his text, the influence of maritime power in history, published at the end of the nineteenth century expressed "those who ignore the lessons of history, are destined to repeat their mistakes". A scenario that Colombia presents today and its relationship with the use of its seas, where the largest settlements of poverty in the country, e.g., the Pacific and the upper Guajira are located on its coasts (Tassin 2016).

Current Situation of the Colombian

The maritime borders of Colombia have an approximate extension of 3,000 thousand kilometers with an underutilization of resources that is close to 100% regarding the condition to the condition of the country because the real exploitation has been undertaken by logistics groups, maritime and port organizations that have been in charge of the exploitation for the last 26 years. They have been the greatest beneficiaries of the natural conditions of the country with respect to the oceans. It is unfortunate to perceive the deep contrast that occurs, for instance, in Buenaventura. Here, poverty is above 90% while the great benefits of port infrastructure disappear from the country while generating a social gap of great dimensions in the territory. A similar situation occurs in the cities of Cartagena, Santa Marta and Barranquilla with the National Conglomerate of Maritime and Inland Harbors and Ports known as 'The Port Society', and multinational companies that are strategically located on our coasts, taking the best and most advantages of our geostrategic location.

Notwithstanding, the aforementioned facts, the sea is used to bring in weapons with which rampant violence is put into practice in our society; our ports and harbors are used for international drug trafficking, smuggling and other irregular activities that are facilitated across the seas. While all this is happening, social conditions such as poverty and misery are of higher proportions to the point that in the Choco and upper Guajira coasts, the highest infant mortality rate is due to lack of socio-economic assistance, and mainly food shortage is reported.

With this document, it is feasible to present a proposal that will help regenerate a better connection with the Colombian central State government. Thus, it must have a realistic development and improvement of its maritime potential. Our government has not even been or heeded or complied with the legally mandatory assistance ordered by the judicial system. International legal intervention and trials against our country that affect our economy like the case in The Hague on November 19, 2012, where our country lost more than 200 nautical miles of coastal waters and territory to Nicaragua when the International Jury voted against Colombia (Guerrero 2017).

Economic Analysis of the Volume of Movement of TEUs in Colombia With Respect to the Countries Involved in the Study (2017–2018)

The study carried out above of the maritime traffic of containers in the world is a benchmarking type reference for the Colombian State. Because it has 3,000 km of maritime borders and is located on the margins of the Panama Canal and it is divided into two oceans, there is a clear indication of the underutilization of the sea. However, to the progressive growth of the movement of TEUs in 1994 when the Sociedad Portuaria de Cartagena, Santa Marta, Barranquilla and Buenaventura were born up to the present (see Table 3). Ports are currently being built in the Urabá area of Antioquia, connected to fourth-generation highways in the city of Medellin and the department of Antioquia, which being in this region of the strongest national development, will surely contribute to the increase in the movement of TEUs, a project that plans to start operations in the second half of 2022 (Kent and Hochstein 1998).

Table 3. Port movement in TEUs by port and port area in Colombia. Source: Economic Commission for Latin America and the Caribbean. Recovered from: https://www.cepal.org/es/notas/informe-la-aactivity-portuaria-america-latina-caribe-2018.

Item	País	Name of the port and port area	Throughput (TEU) 2018	Throughput (TEU) 2017
1	Colombia	Cartagena Bay	2,862,787	2,678,005
2	Colombia	Buenaventura	1,369,139	920,000
3	Colombia	Barranquilla port area	154,533	150,000
4	Colombia	Santa Marta port area	104,611	120,000
5	Colombia	Turbo	73,328	67,522
6	Colombia	San Andrés	15,599	8,134
7	Colombia	Guajira	2,716	1,798

Conclusions

Challenges and Threats for Colombia

According to the pronouncement of the United Nations, "education is the basis of the development of society"; it is interpreted that as educated people do not let themselves be dominated. In addition, education is objective and transparent when the human being is empowered to face with determination the needs of the social construction and generates alternative solutions that contribute to the progress and development of the different scenarios.

That is why there are two major projects in the city of Cartagena, which emerged from the Almirante Padilla Naval Cadets School. They are an icon of the relative maritime underdevelopment that the Colombian state possesses; these are the Cartagena and Cotecmar Regional Port Society 'Corporation of Science and Technology for the Development of the Maritime and River Naval Industry'. If the generation of the mentioned projects were possible, one from the point of view of the mobility of goods that feed international business and with it the foreign trade, whereas others not only potentiate the repair assistance and maintenance of motor vessels but also in its development said project designs and builds vessels, objective evidence of being possible to generate and innovate large-scale projects in the maritime scenarios.

The social debt of the Colombian state with the coastal areas is in the absence of projects that contribute to the transformation of these communities based on the exploration and exploitation of marine resources as an element that contributes to the Nation's GDP.

In light of the circumstances, it is honestly proposed that state policies be generated that contribute to:

- Potentiate the legal scheme, concerning strengthening powers that promulgate with Colombia's borders and limits in the oceans.
- Implement contributory strategies in strengthening maritime and/or coastal culture in history chairs, consolidating in the areas of influence, that is, in the coastal and coastal towns.
- Develop projects for the exploration and exploitation of maritime resources, strengthening competencies in the said process through the Institution of Education, National Learning Service, Sena.
- Strengthening of environmental chairs in an effort to raise awareness about the healthy and objective exploitation of natural resources, keeping in mind the conservation and preservation of biodiversity and ecosystems.
- Strengthen the technical average in coastal areas, strengthening the exploration and exploitation of natural resources found in the seas.
- Encourage all scenarios that explore and exploit natural resources in marketing dynamics in national and international consumer markets.

References

Ahmed, S. 2015. A general review of the history of china's sea-power theory development. Naval War College Review, 68(4), 80–93.

Andrade, T. 2017. The Gunpowder Age: China, Military Innovation, and the Rise of the West in World History. Princeton University Press.

Aubet, M. E. 2008. Political and economic implications of the new Phoenician chronologies. Beyond the Homeland: Markers in Phoenician Chronology, 247–259.

Bridge, C. 2001. Australia's and Canada's Wars, 1914–1918 and 1939–1945 Some Reflections. The Round Table, 90(361), 623–631.

Eubank, M., Holder, T., Lowry, R., Manley, A., Maynard, I., McCormick, A. and Lafferty, M. 2019. All roads lead to Rome, but Rome wasn't built in a day. Advice on QSEP navigation from the 'Roman Gods' of assessment. Sport and Exercise Psychology Review, 15(2).
Fairbanks, M. and Lindsay, S. 1997. Plowing the sea: nurturing the hidden sources of growth in the developing world. Harvard Business School Press.
Fiensy, D. A. and Hawkins, R. K. (eds.). 2013. The Galilean Economy in the Time of Jesus (Vol. 11). SBL Press.
Fuhrmann, C. J. and Fuhrmann, C. 2012. Policing the Roman Empire: Soldiers, administration, and public order. OUP USA.
Garrison, V. H., Shinn, E. A., Foreman, W. T., Griffin, D. W., Holmes, C. W., Kellogg, C. A. and Smith, G. W. 2003. African and Asian dust: from desert soils to coral reefs. BioScience, 53(5), 469–480.
Guerrero Castro, J. E. 2017. Maritime interdiction in the war on drugs in Colombia: practices, technologies and technological innovation.
Hoffmann, J. and Kumar, S. 2010. Globalisation—the maritime nexus. The Handbook of Maritime Economics and Business, 35–64.
Kent, P. E. and Hochstein, A. 1998. Port reform and privatization in conditions of limited competition: the experience in Colombia, Costa Rica and Nicaragua. Maritime Policy & Management, 25(4), 313–333.
Lloyd, B. and Naisbitt, J. 1994. Megatrends and global paradoxes. Management Decision.
Mahan, A. T. 2011. The influence of sea power upon history, 1660–1783. Read Books Ltd.
Mastanduno, M. 1997. Preserving the unipolar moment: Realist theories and U.S. Grand strategy after the cold war. International Security, 21(4), 49–88. doi:10.2307/2539283.
Mirzoyan, A. 2010. Rusia: The Indispensable Ally? In Armenia, the Regional Powers, and the West. pp. 21–53. Palgrave Macmillan, New York.
Oosterhoff, R. 2012. Early modern mathematical practice in the round between Raphael and Galileo: Mutio Oddi and the mathematical culture of late Renaissance Italy. Alexander Marr. Chicago University Press, Chicago, 2011, pp. 384, ISBN 978-0-2265-0628-9.
Rivarola Puntigliano, A. 2011. 'Geopolitics of integration' and the imagination of South America. Geopolitics, 16(4), 846–864.
Scarfi, J. P. 2016. In the name of the Americas: The Pan-American redefinition of the monroe doctrine and the emerging language of american international law in the western hemisphere, 1898–1933. Diplomatic History, 40(2), 189–218.
Spivey, N. 2015. Classical Civilization: A History in Ten Chapters. Head of Zeus.
Stokes, D. and Waterman, K. 2017. Security leverage, structural power and US strategy in East Asia. International Affairs, 93(5), 1039–1060.
Tassin, V. J. 2016. 19 Territorial and Maritime Dispute (Nicaragua v. Colombia), 2001. Latin America and the International Court of Justice: Contributions to International Law, 225.
Tsegay, S. M. (2016). Analysis of globalization, the planet and education. International Journal of Environmental and Science Education, 11(18), 11979–11991.
Ventrella, S. W. 2012. The Power of Positive Thinking in Business: 10 Traits for Maximum Results. Random House.
Wang, N., Zhu, H., Guo, Y. and Peng, C. 2018. The heterogeneous effect of democracy, political globalization, and urbanization on PM2. 5 concentrations in G20 countries: Evidence from panel quantile regression. Journal of Cleaner Production, 194, 54–68.
White, J. A. 2015. Diplomacy of the Russo-Japanese War. Princeton University Press.
Xiang, H. 2014. The Developmental State with Chinese Characteristics under the Neoli-Beralism. Lecture Notes in Management Science Volume 31, 71.

Index

R

S